Imbedding Methods
in Applied Mathematics

Applied Mathematics and Computation

A Series of Graduate Textbooks, Monographs, Reference Works

Series Editor: ROBERT KALABA, University of Southern California

No. 1   MELVIN R. SCOTT
   Invariant Imbedding and Its Applications to Ordinary Differential Equations: An Introduction, 1973

No. 2   JOHN CASTI and ROBERT KALABA
   Imbedding Methods in Applied Mathematics, 1973

In Preparation:

No. 3   DONALD GREENSPAN
   Discrete Models

Imbedding Methods
in Applied Mathematics

JOHN CASTI
University of Arizona

ROBERT KALABA
University of Southern California

 1973

Addison-Wesley Publishing Company
Advanced Book Program
Reading, Massachusetts

London · Amsterdam · Don Mills, Ontario · Sydney · Tokyo

226267

Library of Congress Cataloging in Publication Data
Casti, J   L
   Imbedding methods in applied mathematics.
   (Applied mathematics and computation no. 2)
   Includes bibliographical references.
   1. Electronic data processing—Boundary value
problems. 2. Electronic data processing—Initial
value problems. 3. Invariant imbedding.
I. Kalaba, Robert E., joint author. II. Title.
QA379.C37   515'.62   73–6977
ISBN 0–201–00918–8
ISBN 0–201–00919–6 (pbk.)

Reproduced by Addison-Wesley Publishing Company, Inc., Advanced Book Program, Reading, Massachusetts, from camera-ready copy prepared by the authors.

American Mathematical Society (MOS) Subject Classification Scheme (1970): 47A55, 45B05, 65L10, 65R05

Manufactured in the United States of America
ISBN 0-201-00918-8 (hardbound)
ISBN 0-201-00919-6 (paperback)
ABCDEFGHIJ-MA-7876543

CONTENTS

CHAPTER III. TWO-POINT BOUNDARY VALUE PROBLEMS (continued)

CHAPTER IV. FREDHOLM INTEGRAL EQUATIONS 105

Execution times of modern digital computers are measured in nanoseconds. They can solve hundreds of simultaneous ordinary differential equations with speed and accuracy. But what does this immense capability imply with regard to solving the scientific, engineering, economic, and social problems confronting mankind? Clearly, much effort has to be expended in finding answers to that question.

In some fields, it is not yet possible to write mathematical equations which accurately describe processes of interest. Here, the computer may be used simply to simulate a process and, perhaps, to observe the efficacy of different control processes. In others, a mathematical description may be available, but the equations are frequently difficult to solve numerically. In such cases, the difficulties may be faced squarely and possibly overcome; alternatively, formulations may be sought which are more compatible with the inherent capabilities of computers. Mathematics itself nourishes and is nourished by such developments.

Each order of magnitude increase in speed and memory size of computers requires a reexamination of computational techniques and an assessment of the new problems which may be brought within the realm of solution. Volumes in this series will provide indications of current thinking regarding

problem formulations, mathematical analysis, and computational

treatment.

                                            ROBERT KALABA

Los Angeles, California
April, 1973

PREFACE

According to Von Neumann, "mathematical ideas origi-
nate in empirics". As a result of stimuli from the physical
world, mathematical approximations to reality are made and
new theories created. Those theories that flourish and enter
the mainstream of mathematical thought are the ones which are
capable of bringing together significant classes of problems
under a common framework and of answering previous unasked
questions. These two features of universality and conceptual
illumination distinguish a mathematical theory from an ad hoc
technique. One of the most effective theory-producing stimuli
in the history of mathematics has been the pressure of obtaining numerical answers to numerical questions. It can safely
be said that a large part of what is now called classical
applied mathematics was developed in an attempt to avoid
doing arithmetic. In modern times, following the development
of the electronic computer, the mathematical emphasis has
shifted from the complete avoidance of arithmetic to the
carrying out of arithmetical operations in the most effective
and efficient manner.

The primary idea motivating the work of this book is
that digital computers, by virtue of their iterative nature,
are the most effective means available for the solution of
differential equations whose data are all specified at a
single point, a so-called "initial value problem". On the
other hand, a large fraction of the problems encountered in
mathematical physics, engineering, biology, economics, and
operations research are of a boundary value nature, i.e., the
conditions necessary to specify a solution are given at two
or more points. Suffice it to say that digital computers are
not, in general, well-suited for the numerical solution of
this type of problem. The theme of the book is that the
mathematical formulation of a problem should be chosen keeping
in mind the computational tools available and that often
alternate, but equivalent, formulations exist which may possess
distinct advantages not available in the original setting.
Too often the traditional mathematical formulation of a problem
is accepted on faith as the formulation, as if it had been
prescribed by a mathematical Solomon whose wisdom, judgment,
and foresight were beyond question.

Making use of a mathematical idea which had its
origins in the transfer of radiation through the atmosphere,
our aim is to show how the initial fruits of the mathematical
theory of invariant imbedding may be employed to recast broad
classes of boundary value problems as initial value problems.
In addition to our main objective of supplying the working

xi

scientist with a potent tool to add to his arsenal of weapons
for attacking important problems in his field, we expose a
chain of ideas which appear to have the potential of developing
into one of the mainstreams of modern applied mathematics.

In order to make the book accessible to as wide an
audience as possible, we have deliberately adopted a mathematically formal, rather than rigorous, presentation. While
"purists" may look rather askance at this approach, we think
that a formal development, interspersed with numerous examples,
is the best vehicle with which to isolate the basic ideas
involved, unencumbered by artificially erected mathematical
roadblocks and detours which are inevitably encountered in a
rigorous development. Needless to say, we recognize that the
validity of our ideas depends, to a large degree, upon estab-
lishing them on a firm mathematical foundation. However, we
prefer to relegate all but the most essential theorems to a
future volume. We hope that this policy will not only make
the current volume accessible to the general scientific commu-
nity, but will also serve to catalyze further theoretical
developments by others of the mathematical fraternity. It is
our belief that numerous journal articles, theses, and books
are waiting to be uncovered just below the surface of formalism
presented here. In any event, the reader with an undergraduate
course in ordinary differential equations should have little
difficulty following the majority of our lines of thought.
Let us now briefly describe the scope and content of the book.

In the first chapter, the basic ideas we employ are
introduced through the medium of finite difference equations.
A brief discussion of the "gambler's ruin" problem is given
and an alternate to the classical formulation is presented
which is shown to possess computational (and analytic) advantages over the usual formulation. Following this example, we
present our ideas for a general inhomogeneous linear difference
system.

To make the book as self-contained as possible, in
Chapter Two we present a brief overview of the philosophy and
methodology for the numerical solution of initial value
problems. Standard numerical integration schemes such as
Runge-Kutta and Adams-Moulton are treated, as well as questions
of numerical stability.

The major thrust of our development begins in Chapter
Three with an investigation of linear and nonlinear two-point
boundary value problems. In addition to showing that initial

value representations exist for a broad class of problems, we
also present results suitable for obtaining Green's functions.
The numerical stability of the initial value procedures are
briefly treated through an example, and we present the results
of several numerical experiments to validate the methods
proposed.

In Chapter Four the topic of interest is Fredholm
integral equations. For the most part, we deal with equations
having a displacement-type kernel, i.e., $k(t, y) = k(|t - y|)$.
The reason for restricting the majority of our attention to
this type of problem is that such equations occur with great
frequency in many branches of physics, astronomy, engineering,
and biology and, as might be expected, taking advantage of
the special structure of the kernel allows us to obtain more
precise results than in the general case. References are given
at the end of the chapter to similar treatments of more general
kernels of degenerate and semi-degenerate type.

Variational theory and control problems occupy our
attention in Chapter Five. It is shown how significant classes
of constrained and unconstrained optimization problems may be
attacked from a point of view independent of Euler equations,
dynamic programming, or the maximum principle. However, we
also point out by way of example that a combination of methods
may be expected to produce a solution procedure superior to
that obtained from a rigid adherence to any one particular
method.

The last chapter is devoted to an application of the
ideas developed in the foregoing chapters to the physical
sciences. It is shown how use of the imbedding ideas leads to
new formulations and insights into problems arising in radiative transfer, analytical mechanics, optimal filtering theory,
and non-local wave interaction.

As is evident from a glance at the references in each
chapter, much of the work reported in this book was carried
out in collaboration with others. So, in a very real sense
this is as much their book as ours. In this regard, we
particularly wish to acknowledge conversations and collabora-
tions with S. Ueno and H. Kagiwada on topics in radiative
transfer, R. Sridhar for fruitful discussions on filtering
and estimation theory, A. Schumitzky for many helpful suggestions on integral equations, E. Angel and M. Juncosa for
critical comments and suggestions on numerical questions,
M. Scott for help with characteristic value problems, and

above all to R. Bellman whose influence is felt in all the
topics discussed here, and without whose interest, encourage-
ment, and enthusiasm, both the concepts of invariant imbedding
and this book would have remained in the realm of undreamt
dreams.

In addition, a vote of thanks is due to a myriad of
typists at the RAND Corporation, Systems Control, Inc., and
the University of Southern California for unflinchingly typing
(and re-typing) the various sections of the book. In particular, we wish to single out Kanda Kunze who integrated the
original patchwork version of this book into a polished final
manuscript.

March, 1973
Tucson, Arizona   John Casti
Los Angeles, California   Robert Kalaba

CHAPTER ONE

FINITE DIFFERENCE EQUATIONS

1. INTRODUCTION

The theory of invariant imbedding requires us to re-orient our thinking about how mathematical problems should be posed. A formulation which is satisfactory for analytic investigations may be useless computationally, and vice versa. Witness Cramer's rule for the solution of linear algebraic equations. In this chapter we wish to examine a simple physical process, first from the traditional viewpoint and then from the standpoint of invariant imbedding. In this manner, we are able to point out the characteristic features of both approaches and to examine the merits of each.

The problem we wish to consider is the classical "gambler's ruin". In spite of its conceptual simplicity, this problem contains all the features necessary to exhibit

1

the invariant imbedding concept. Complete understanding of

this example will serve as a stepping-stone to the more com-

plex examples treated in subsequent chapters.

Following our discussion of the physical process, we

take up the treatment of the more general problem of

invariant imbedding and a system of linear difference equa-

tions. Besides its intrinsic interest, this treatment serves

as a model for what is done in Chapter Two for two-point

boundary value problems.

2. THE "GAMBLER'S RUIN"

Consider a gambler who possesses a finite fortune k.

Assume that he engages in a game in which there is a proba-

bility $p(k)$ of increasing his fortune to $(k + 1)$ and a

probability $q(k) = 1 - p(k)$ of decreasing to $(k - 1)$.

Furthermore, let us assume that he agrees to play until he

either amasses a fortune N or loses all his money. The

question is: Given the probabilities $\{p(k)\}$, $k = 1,2,\ldots,$

$N - 1$, what is the probability that our gambler will attain

his objective before going broke? The reader will recognize

this as an inhomogeneous version of the classical gambler's

ruin in which the $\{p(k)\}$ are constant.

3.   CLASSICAL FORMULATION

The usual formulation of the gambler's ruin problem
proceeds by introducing the function u(k) defined as

u(k, N) = the probability that the gambler

attains a fortune N before going

broke when he starts with a fortune k.

Translating the verbal description of the problem into a
mathematical equation for the function u we have the difference equation

$$u(k, N) = p(k)u(k + 1, N) + q(k)u(k - 1, N)$$
$$k = 1,2,\ldots,N - 1, N \geq 2. \tag{1}$$

The conditions at k = 0 and k = N are

$$u(0, N) = 0, \tag{2}$$

$$u(N, N) = 1, \tag{3}$$

expressing the fact that with no capital the gambler cannot
play and with N units of capital he has already achieved
his objective. We write u(k, N) to explicitly indicate
that u depends upon the amount the gambler wishes to
amass before stopping.

In spite of its simple appearance as a linear

second-order difference equation, Eq. (1)[1] possesses, in

general, no explicit solution for non-constant p and q.

Consequently, recourse must be taken to numerical methods.

We see that if conditions had been prescribed at two adjacent

points k = n,n - 1, n = 1,2,...,N, we could use the pre-

scribed values together with Eq. (1) to produce the function

u(k) in a straightforward iterative manner. This is the

type of computation with which digital computers are most

happy. Unfortunately, the conditions (2) and (3) are not

prescribed in such a favorable fashion, being given at the

separated points k = 0 and k = N. The result of this

"two-point" boundary condition is that no routine iterative

computational methods can be employed. Let us see how an

alternative way of viewing the problem offers a way out of

this analytic and computational impasse.

4.   INVARIANT IMBEDDING APPROACH

Consider the same process as before except that now

the gambler desires to attain an amount N + 1 before

---

[1]
 Throughout this book, references to equations shall have the
following interpretation: a single equation number such as
Eq. (1), will refer to the corresponding equation within the
section in which the reference occurs. A double number, e.g.
Eq. (5.3), refers to the third equation within section five
of the chapter.

stopping. The appropriate boundary value problem is now

$$u(k, \ N + 1) = p(k)u(k + 1, \ N + 1) +$$

$$q(k)u(k - 1, \ N + 1), \quad (1)$$

$$k = 1,2,\ldots,N,$$

$$u(0, \ N + 1) = 0, \tag{2}$$

$$u(N + 1, \ N + 1) = 1. \tag{3}$$

We wish to obtain a relationship between the two problems
(3.1)-(3.3) and (1)-(3). To do this we observe that in order
to attain a fortune $N + 1$, the gambler must first attain a
fortune N without going broke. This argument leads to the
mathematical relation

$$u(k, \ N + 1) = u(k, \ N)u(N, \ N + 1), \tag{4}$$

$$N \geq k.$$

To make use of Eq. (4) we need an expression for
$u(N, \ N + 1)$. Putting $k = N$ in Eq. (1) gives

$$u(N, \ N + 1) = p(N)u(N + 1, \ N + 1) +$$

$$q(N)u(N - 1, \ N + 1) \tag{5}$$

which, taking account of Eq. (3), yields

$$u(N, \ N + 1) = p(N) + q(N)u(N - 1, \ N + 1). \tag{6}$$

Arguing as before, we see that

$$u(N - 1, N + 1) = u(N - 1, N)u(N, N + 1). \qquad (7)$$

Introducing the function r defined by

$$r(N) = u(N, N + 1), \qquad N \geq 0, \qquad (8)$$

Eq. (6) is rewritten as

$$r(N) = p(N) + q(N)[r(N - 1)r(N)]. \qquad (9)$$

Solving Eq. (9) for $r(N)$ gives

$$r(N) = \frac{p(N)}{1 - q(N)r(N-1)}, \qquad N = 1, 2, \ldots . \qquad (10)$$

The initial condition at $N = 0$ is

$$r(0) = 0, \qquad (11)$$

expressing the fact that the gambler has no chance of reaching his goal if he begins with no capital.

Returning to Eq. (4), we may now write

$$u(k, N + 1) = u(k, N)r(N), \qquad N \geq k, \qquad (12)$$

where the function r is produced by means of Eqs. (10) and (11). The initial condition on u is

$$u(k, k) = 1. \qquad (13)$$

Observe that Eqs. (10)-(13) for u and r define an initial value problem in contrast to the two-point boundary value problem of the classical formulation. Consequently, for computational purposes the nonlinear invariant imbedding formulation may be preferable to the linear classical problem.

The computational procedure to determine the function $u(k, N)$ for any fixed k, $0 < k < N$, is the following: Use Eqs. (10) and (11) to produce the values $r(1),\ldots,r(k)$. At $N = k$, adjoin Eq. (12) with initial condition $u(k, k) = 1$. For $N > k$, continue producing the function r and use r to obtain $u(k, N + 1)$ by Eq. (12). The process stops when N reaches the desired fortune, say N_0. At this point the number $u(k, N_0)$ is available giving the probability that the gambler will achieve a fortune N_0 before going broke when he starts with an amount k. Note that the solution may be obtained for any value of k by just adjoining an additional equation of the form of Eq. (12) as N reaches the particular value of k desired.

5. A STANDARD EXAMPLE

To show that the invariant imbedding equations are in complete agreement with the results of classical theory, let us examine the case where p and q are constant. In this case the closed form solution of the linear boundary value problem (3.1)-(3.3) is

$$u(k, N) = 1 - \frac{\left(\frac{q}{p}\right)^N - \left(\frac{q}{p}\right)^k}{\left(\frac{q}{p}\right)^N - 1} , \qquad p \neq q \qquad (1)$$

$$= \frac{k}{N} , \qquad p = q, \quad k \leq N. \qquad (2)$$

We shall verify that the functions r and u of Eqs. (4.10)-(4.13) satisfy the Eq. (2), leaving the more general case to the reader.

To verify Eq. (4.10), we must show that

$$u(N, N + 1) = \frac{p}{1 - qu(N-1,N)} , \qquad (3)$$

and

$$u(0, 1) = 0. \qquad (4)$$

Eq. (4) is obviously satisfied by Eq. (2). To check Eq. (3), we write

$$\frac{p}{1 - qu(N-1,N)} = 1 - q\left(\frac{N-1}{N}\right)$$

$$= \frac{\frac{1}{2}}{1 - \frac{1}{2} \frac{N-1}{N}}$$

$$= \frac{N}{N+1}$$

$$= u(N, N + 1). \tag{5}$$

To establish Eqs. (4.12) and (4.13), we note that Eq. (2) gives

$$u(k, k) = 1, \quad k = 0, 1, \ldots, N \tag{6}$$

which is Eq. (4.13). Now write

$$u(k, N)r(N) = \frac{kN}{N(N+1)}$$

$$= \frac{k}{N+1}$$

$$= u(k, N + 1).$$

This is Eq. (4.12).

6. LINEAR DIFFERENCE EQUATIONS

Having introduced our basic ideas by means of simple physical arguments, let us now consider the problem from a purely mathematical point of view laying aside all physical models.

Consider the linear homogeneous difference equations

$$u_{n+1} = Au_n + Bv_n, \qquad u_0 = 0, \tag{1}$$

$$v_{n+1} = Cu_n + Dv_n, \qquad v_N = 1, \tag{2}$$

$$n = 1,2,\ldots,N - 1.$$

The usual method for converting this boundary value problem into an initial value system is to hold the interval length N fixed. We then write

$$u_n = \rho_n v_n, \qquad n = 0,1,\ldots,N \tag{3}$$

and

$$\rho_{n+1}v_{n+1} = A\rho_n v_n + Bv_n. \tag{4}$$

$$\rho_{n+1}\left[C\rho_n v_n + Dv_n\right] = A\rho_n v_n + Bv_n. \tag{5}$$

Solving for ρ_{n+1} gives

$$\rho_{n+1} = \frac{A\rho_n + B}{C\rho_n + D}, \qquad n = 0,1,\ldots,N - 1, \tag{6}$$

a linear fractional recurrence relation having the initial condition

$$\rho_0 = 0. \tag{7}$$

Using Eqs. (6) and (7), the sequence $\rho_0, \rho_1, \ldots, \rho_N$ is readily determined. In particular, the missing boundary condition ρ_N is determined. Assuming $AD - BC \neq 0$, the variables u_{N-1}, v_{N-1} can be found by solving the set of linear algebraic equations

$$u_N = Au_{N-1} + Bv_{N-1}, \tag{8}$$

$$v_N = Cu_{N-1} + Dv_{N-1}. \tag{9}$$

Then knowing the values of u_{N-1} and v_{N-1}, the values of u_{N-2} and v_{N-2} are determined, and so on.

This classical approach is a two-sweep method. The first involves determining the sequence $\{\rho_n\}$, the second involves determining the pairs $\{(u_n, v_n)\}$.

7. INVARIANT IMBEDDING

The invariant imbedding approach to the solution of Eqs. (1) and (2) is conceptually different. The parameter N is varied, while n is held fixed. As we shall see, this mode of attack enables us to simultaneously bypass boundary-value problems and two-sweep methods.

Since the solution of Eqs. (6.1) and (6.2) depends upon both n and N, let us adopt a notation which emphasizes this. We write the boundary value problem in the form

$$u(n + 1, N) = Au(n, N) + Bv(n, N), \quad u(0, N) = 0, \quad (1)$$

$$v(n + 1, N) = Cu(n, N) + Dv(n, N), \quad v(N, N) = 1,$$
$$n = 0,1,2,\ldots,N - 1. \quad (2)$$

Consider the related problem on the interval of length $N + 1$,

$$u(n + 1, N + 1) = Au(n, N + 1) +$$
$$Bv(n, N + 1), \quad u(0, N + 1) = 0, \quad (3)$$

$$v(n + 1, N + 1) = Cu(n, N + 1) +$$
$$Dv(n, N + 1), \quad v(N + 1, N + 1) = 1. \quad (4)$$

In terms of the process on an interval of length N, we have the same system as that in Eqs. (6.1) and (6.2) except that the boundary condition on v is now $v(N, N + 1)$ instead of 1. Thus, assuming that the two linear problems (1)-(2) and (3)-(4) have unique solutions, we see that they are related by the equations

$$u(n, N + 1) = v(N, N + 1)u(n, N), \quad (5)$$

$$v(n, N + 1) = v(N, N + 1)v(n, N), \quad N \geq n. \quad (6)$$

To make use of these relations we shall need initial conditions on u and v at N = n, and we shall need more detailed information about the quantity v(N, N + 1) as a function of N.

In Eqs. (3) and (4) put n = N. This gives

$$r(N + 1) = Au(N, N + 1) + Bv(N, N + 1), \tag{7}$$

$$1 = Cu(N, N + 1) + Dv(N, N + 1), \tag{8}$$

where we have introduced

$$r(N) = u(N, N), \qquad N = 0,1,2,\ldots . \tag{9}$$

Assuming that $\Delta = AD - BC \neq 0$, Eqs. (7) and (8) may be solved in the form

$$u(N, N + 1) = \frac{Dr(N+1) - B}{\Delta} \tag{10}$$

$$v(N, N + 1) = \frac{A - Cr(N+1)}{\Delta} ,$$

$$N = 0,1,2,\ldots . \tag{11}$$

We now derive the recurrence relations for determining the sequence $\{r(N)\}$. In Eq. (5) put n = N, and use Eqs. (9)-(11) to obtain

$$\frac{Dr(N+1) - B}{\Delta} = \frac{A - Cr(N+1)}{\Delta} \, r(N). \qquad (12)$$

Solving for $r(N + 1)$ yields the recurrence relation

$$r(N + 1) = \frac{Ar(N) + B}{Cr(N) + D}, \qquad N = 0,1,2,\ldots . \qquad (13)$$

The initial condition at $N = 0$ is

$$r(0) = 0. \qquad (14)$$

Comparing the Cauchy systems for the sequences $\{r(i)\}$, we see that

$$\rho_i = r(i), \qquad i = 0,1,2,\ldots . \qquad (15)$$

To this extent the methods overlap. The defining relations and modes of reasoning, through, are different.

8.   THE INITIAL VALUE SYSTEM

Let us now state the one sweep method of invariant imbedding for determining the solution of Eqs. (6.1) and (6.2). Say we want the solution at a fixed positive integer n. Using the initial condition

$$r(0) = 0 \qquad (1)$$

and the recurrence relation

$$r(N + 1) = \frac{Ar(N) + B}{Cr(N) + D} , \tag{2}$$

determine $r(1), r(2), \ldots, r(n)$. Then adjoin the recurrence relations

$$u(n, N + 1) = \frac{A - Cr(n+1)}{\Delta} u(n, N), \tag{3}$$

$$v(n, N + 1) = \frac{A - Cr(n+1)}{\Delta} v(n, N), \qquad n \geq N. \tag{4}$$

As initial conditions at $N = n$, use

$$u(n, n) = r(n), \tag{5}$$

$$v(n, n) = 1. \tag{6}$$

Since the number $r(n)$ has been calculated, these are known initial conditions. In this manner the triples $[r(n + 1), u(n, n + 1), v(n, n + 1)], \ldots$, are determined.

9. THE INHOMOGENEOUS CASE

Consider the linear inhomogeneous difference equations

$$x(n + 1, N) = Ax(n, N) + By(n, N) + f(n), \tag{1}$$

$$y(n + 1, N) = Cx(n, N) + Dy(n, N) + g(n),$$

$$n = 0,1,2,\ldots,N - 1, \tag{2}$$

subject to the inhomogeneous boundary conditions

$$x(0, N) = 0, \tag{3}$$

$$y(N, N) = 1. \tag{4}$$

The solution of this system may be written in the form

$$x(n, N) = p(n, N) + u(n, N), \tag{5}$$

$$y(n, N) = q(n, N) + v(n, N), \qquad 0 \le n \le N. \tag{6}$$

The functions $p(n, N)$ and $q(n, N)$ satisfy the inhomogeneous difference equations

$$p(n + 1, N) = Ap(n, N) + Bq(n, N) + f(n), \tag{7}$$

$$q(n + 1, N) = Cp(n, N) + Dq(n, N) + g(n),$$

$$n = 0,1,2,\ldots,N - 1, \tag{8}$$

and the homogeneous boundary conditions

$$p(0, N) = 0, \tag{9}$$

$$q(N, N) = 0. \tag{10}$$

Since we have already shown how to determine the functions

$u(n, N)$ and $v(n, N)$, $N \geq n$, as solutions of a Cauchy problem, we have merely to carry out the derivation for the functions $p(n, N)$ and $q(n, N)$, $N \geq n$.

Consider the corresponding problem on the interval of length $N + 1$. The equations are

$$p(n + 1, N + 1) = Ap(n, N + 1) + Bq(n, N + 1)$$
$$+ f(n), \quad (11)$$

$$q(n + 1, N + 1) = Cp(n, N + 1) + Dq(n, N + 1)$$
$$+ g(n), \quad 0 \leq n \leq N, \quad (12)$$

$$p(0, N + 1) = 0, \quad (13)$$

$$q(N + 1, N + 1) = 0. \quad (14)$$

The pair of differences, $p(n, N + 1) - p(n, N)$ and $q(n, N + 1) - q(n, N)$, satisfy homogeneous difference equations for $0 \leq n \leq N$. At $n = 0$ the first difference is zero, and at $n = N$ the second difference is $q(N, N + 1)$. It follows that

$$p(n, N + 1) - p(n, N) = q(N, N + 1)u(n, N), \quad (15)$$

$$q(n, N + 1) - q(n, N) = q(N, N + 1)v(n, N),$$
$$0 \leq n \leq n. \quad (16)$$

These are the basic recurrence relations for calculating the functions p and q for $N \geq n$. Since the functions u and v are already characterized as the solution of a Cauchy system, we have but to consider the factor $q(N, N + 1)$, $N = 0,1,2,\ldots$.

In Eqs. (11) and (12) put $n = N$. The result is that

$$S(N + 1) = Ap(N, N + 1) + Bq(N, N + 1) + f(N), \quad (17)$$

and

$$0 = Cp(N, N + 1) + Dq(N, N + 1) + g(N), \quad (18)$$

where

$$S(N) = p(N, N). \quad (19)$$

The solution of Eqs. (17) and (18) is

$$p(N, N + 1) = \frac{D[S(N+1) - f(N)] + Bg(N)}{\Delta}, \quad (20)$$

$$q(N, N + 1) = \frac{-Ag(N) - C[S(N+1) - f(N)]}{\Delta}. \quad (21)$$

Putting $n = N$ in Eq. (15) yields the equation

$$p(N, N + 1) - p(N, N) = q(N, N + 1)u(N, N). \quad (22)$$

The last relation may be rewritten as

$$\frac{D[S(N+1) - f(N)] + Bg(N)}{\Delta} - S(N)$$

$$= r(N) \; \frac{-Ag(N) - C[S(N+1) - f(N)]}{\Delta} \, . \qquad (23)$$

Solving for $S(N + 1)$, we obtain the desired recurrence relation

$$[D + Cr(N)]S(N + 1) = -[B + Ar(N)]g(N)$$
$$+ \; Cf(N)r(N) + S(N), \qquad N = 0,1,2,\ldots \; . \qquad (24)$$

According to the definition in Eq. (19) the initial condition on the function S at $N = 0$ is

$$S(0) = 0. \qquad (25)$$

The function S is determined by the Eqs. (25) and (24). The function $q(N, N + 1)$ is then given by Eq. (21). And finally, the functions p and q are determined by the Eqs. (15) and (16) for $N \geq n$, together with the initial conditions at $N = n$,

$$p(n, n) = S(n), \qquad (26)$$

$$q(n, n) = 0. \qquad (27)$$

CHAPTER ONE

NOTES AND REFERENCES

§2-3.   An elementary, but fairly extensive discussion of the
         "gambler's ruin" problem and its relationship to the
         diffusion equation is given in

> Feller, W., _Introduction to Probability Theory and_
> _its Applications_, 2nd Ed., Vol. 1, John Wiley, New
> York, 1957.

For a more detailed mathematical treatment see

> Spitzer, F., _Principles of Random Walk_, Van
> Nostrand, Princeton, N.J., 1964,

or

> Chung, K., _A Course in Probability Theory_, Harcourt,
> Brace, and World, New York, 1968.

§4.   The first application of invariant imbedding to
         problems of random walk is the paper

> Bellman, R. and R. Kalaba, "Random Walk,
> Scattering, and Invariant Imbedding-I: One-
> Dimensional Discrete Case," _Proc. Nat. Acad. Sci._
> _USA_, Vol. 43, 1957, 930-933.

A more recent treatment is

> Bellman, R., "Invariant Imbedding and Random Walk,"
> _Proc. Am. Mth. Soc._, Vol. 13, 1962, 251-254.

§6.   An elementary exposition of linear difference

equations with numerous applications to economics

and the social sciences is found in

    Goldberg, S., <u>Introduction to Difference Equations</u>,

    John Wiley, New York, 1958.

A more precise mathematical treatment is given by

    Miller, K., <u>Linear Difference Equations</u>, W. A.

    Benjamin Co., New York, 1968.

§7-9.   We follow the development in

    Kalaba, R., "A One-Sweep Method for Linear

    Difference Equations with Two-Point Boundary

    Conditions," USCEE Report 69-23, U. So. Calif.,

    Los Angeles, 1969.

Multidimensional versions of the problem treated here

frequently arise in the solution of partial differ-

ential equations by finite difference schemes. For

a discussion of some of these problems see

    Angel, E. and R. Kalaba, "A One-Sweep Numerical

    Method for Vector-Matrix Difference Equations with

    Two-Point Boundary Conditions," <u>J. Optim. Theory</u>

    <u>and Appl</u>. 1971.

CHAPTER TWO

INITIAL VALUE PROBLEMS

1. INTRODUCTION

In order to keep our discussion self-contained, in this chapter we shall briefly present the rudimentary analytic and computational aspects of initial value problems for ordinary differential equations. Our basic equation will be the nonlinear scalar differential equation

$$u'(t) = h(u(t), t), \quad u(0) = c, \tag{1}$$

although we will occasionally deal with a system of such equations.

Even though the primary aim of this chapter is to acquaint the reader with basic concepts for the numerical solution of initial value problems, the analytic aspects of existence and uniqueness are discussed primarily as a

reminder to the unwary that the computer is an aid to
analysis, not a substitute. The concluding section of this
chapter introduces two-point boundary value problems of the
form

$$u''(t) + a(t)u = 0$$

$$\text{(2)}$$

$$u(0) = c_1, \quad u(T) = c_2$$

A few of the major difficulties associated with the solution
of (2) are presented as motivation for the development of the
methods discussed in the remainder of the book.

2. VECTOR-MATRIX NOTATION

When dealing with systems of differential equations
as we shall do throughout the remainder of this book, in
order to avoid the awkard and cumbersome task of having to
explicitly write out all the equations, we will often employ
vector-matrix notation.

Consider the system of initial value problems

$$\frac{du_1(t)}{dt} = h_1(u_1, u_2, \ldots, u_N, t), \qquad u_1(0) = c_1,$$

$$\frac{du_2(t)}{dt} = h_2(u_1, u_2, \ldots, u_N, t), \qquad u_2(0) = c_2,$$

$$\vdots \qquad\qquad \vdots \qquad\qquad \vdots$$

$$\frac{du_N(t)}{dt} \; = \; h_N(u_1, u_2, \ldots, u_N, t), \qquad u_N(0) = c_N. \qquad (1)$$

Introducing the N-dimensional vectors u, c and h as

$$c = \begin{pmatrix} c_1 \\ c_2 \\ \vdots \\ c_N \end{pmatrix}, \; u = \begin{pmatrix} u_1(t) \\ u_2(t) \\ \vdots \\ u_N(t) \end{pmatrix}; \; h = \begin{pmatrix} h_1(u_1, u_2, \ldots, u_N, t) \\ h_2(u_1, u_2, \ldots, u_N, t) \\ \vdots \\ h_N(u_1, u_2, \ldots, u_N, t) \end{pmatrix},$$

We see that the system (1) may be concisely written as

$$\frac{du}{dt} \; = \; h(u, \; t), \quad u(0) = c, \qquad\qquad\qquad (2)$$

where differentiation of the vector u is defined component-
wise, i.e.,

$$\frac{du}{dt} \; = \; \begin{pmatrix} \dfrac{du_1}{dt} \\ \vdots \\ \dfrac{du_N}{dt} \end{pmatrix}.$$

In the event the system (1) is linear, i.e.,

$$\frac{du_i}{dt} = \sum_{j=1}^{N} a_{ij}(t)u_j(t), \quad u_i(0) = c_i,$$

$$i = 1,2,\ldots,N, \tag{3}$$

the usual notation is

$$\frac{du}{dt} = A(t)u, \quad u(0) = c, \tag{4}$$

where u, $\frac{du}{dt}$, and c are as before and $A(t)$ is the

$N \times N$ matrix

$$A(t) = \begin{pmatrix} a_{11}(t) & \cdots & a_{1N}(t) \\ \vdots & & \\ a_{N1}(t) & \cdots & a_{NN}(t) \end{pmatrix}.$$

3. EXISTENCE AND UNIQUENESS

Before embarking upon our brief journey into the
realm of numerical analysis, let us dwell for a moment upon
two vital analytic questions--the existence and uniqueness of
solutions to initial value problems. It is reasonable to
suppose that before we attempt to produce the numerical solution of a differential equation, it would be desirable to
know whether or not the particular equation possesses a
unique solution. For, if existence fails, any numbers

produced are meaningless and, if the solution is not unique,

care must be taken to determine which solution the computer

is producing. (It is amusing, with a wry twist, to note that

many engineers and physicists dismiss questions of existence

and uniqueness in a rather cavalier manner feeling that only

mathematicians really care about such matters. In actual

fact, the mathematician should be less interested in

guaranteeing existence and uniqueness than the engineer since,

when the engineer writes down a mathematical equation which

is supposed to be a reasonable approximation to a real physi-

cal process, the equation should have a unique solution if it

is to bear any resemblance to reality. On the other hand,

the mathematician finds the situation interesting if exis-

tence and uniqueness occurs and may find it equally interest-

ing if they don't.)

Let us now consider the initial value problem

$$\frac{dz}{dt} = h(z, t), \quad z(0) = c, \tag{1}$$

where   z, h, and   c   are n-dimensional vectors. Let   R   be
the region of R^n defined by $||z - c|| \le k_1$ and let
$k_2 = \max ||h(z, t)||$ for all   z   in   R.   Then we have the
following theorem:

EXISTENCE AND UNIQUENESS THEOREM. If $h(z, t)$ is a continuous function of z in R, and if for any two vectors x
and y in R there exists a constant k_3, independent of
x and y, such that

$$||h(x, t) - h(y, t)|| \leq k_3 ||x - y||, \qquad (2)$$

then (1) has a unique solution for all $0 \leq t < k_1/k_2$.
The proof of this theorem is straightforward and can be found
in the references given at the end of the chapter. It should
be noted that a condition of the type (2) is called a
Lipshitz condition and it is easily verified that it implies
continuity. A simple test on $h(z, t)$ to insure that it
holds is for $h(z, t)$ to have uniformly bounded partial
derivatives with respect to the z_i in R.

4.   NUMERICAL SOLUTION OF INITIAL VALUE PROBLEMS

Let us now turn our attention to the problem of producing the numerical solution of an initial value problem by
means of a digital computer. Say we wish to solve the scalar
equation

$$\frac{du}{dt} = h(u, t), \quad u(0) = c. \qquad (1)$$

Since the digital computer is capable of only the arithmetic

operations of addition, subtraction, multiplication and

division, our first task is to reduce the transcendental

operation of differentiation to an arithmetic one. A simple-

minded, and by no means efficient way is to approximate (1)

by the finite-difference form

$$u(t + \Delta) \overset{\sim}{=} u(t) + h(u,\ t)\Delta, \quad u(0) = c, \tag{2}$$

where Δ is a "small" parameter. Knowledge of the initial

value $u(0) = c$, allows us to use (2) to successively produce

the values $u(\Delta), u(2\Delta), \ldots$ in an iterative fashion, a task

for which a digital computer is admirably suited.

As is always the case when replacing an exact form

by an approximate equation, the question of the accuracy of

the approximation must be investigated. For example, in

(2), upon expanding $u(t + \Delta)$ we have

$$u(t + \Delta) = u(t) + \Delta u'(t) + \frac{\Delta^2}{2} u''(t) + \cdots. \tag{3}$$

Therefore, we see that the local error in using (2) to com-

pute $u(k\Delta)$, given $u((k - 1)\Delta)$, is

$$0(\Delta^2 |u''(k - 1)\Delta|). \tag{4}$$

As an example of what may be done to improve the

estimate (2) with little additional work, we may approximate

u'(t) by central differences which yields the approxima-

tion

$$u(t + \Delta) \overset{\sim}{=} u(t - \Delta) + h(u, t)2\Delta, \qquad u(0) = c. \qquad (5)$$

Expansion of $u(t + \Delta)$ and $u(t - \Delta)$ in powers of Δ about

the point t gives the local error

$$0(\Delta^3 |u'''(t)|), \qquad\qquad\qquad (6)$$

which is one order of magnitude better than that of (4).

The type of approximation formula (2) dates back to

Euler and is usually called Euler's method. Its geometrical

significance is shown in Fig. 1. Starting at the initial

point (0, c), the slope of the function u is calculated

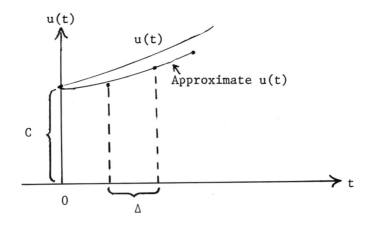

Fig. 1   Illustrations of Euler's Method

and is then used to approximate the function as a straight
line to the point $(\Delta, c + h(c, 0)\Delta)$. The process is then
repeated at the new point in order to obtain the next point
and so on. The polygonal curve connecting the points is the
approximation to $u(t)$.

In order to improve error estimates of Euler's method,
more elaborate recurrence relations must be employed. Among
the most popular of current methods are the various Runge-
Kutta procedures. The characteristic feature of these methods
is that the interval $(t, t + \Delta)$ is broken up into a number
of subintervals, and by a judicious choice of the points
where the function $h(u, t)$ is evaluated, a much higher
order of truncations error is obtained.

The most common Runge-Kutta method is the fourth-
order procedure in which the interval $(t, t + \Delta)$ is broken
up into four sub-intervals. The integration scheme is then
given by the relations

$$k_0 = h(u(t), t)\Delta, \tag{7}$$

$$k_1 = h(u(t) + \beta_{10}k_0, t + \alpha_1\Delta)\Delta, \tag{8}$$

$$k_2 = h(u(t) + \beta_{20}k_0 + \beta_{21}k_1, t + \alpha_2\Delta)\Delta, \tag{9}$$

$$k_3 = h(u(t) + \beta_{30}k_0 + \beta_{31}k_1 + \beta_{32}k_2, t + \alpha_3\Delta)\Delta, \tag{10}$$

$$u(t + \Delta) = u(t) + (ak_0 + bk_1 + ck_2 + dk_3). \tag{11}$$

The numerous parameters appearing in Eqs. (7)-(11) may be partially determined by equating the expression for $u(t + \Delta)$ with the Taylor series expansion of $u(t + \Delta)$ about the point t. Agreement through terms in Δ^4 may be achieved, giving a truncation error of $0(\Delta^5)$. Of the infinitude of possible solutions, the one originally proposed by Runge is

$$k_0 = h(u(t), t)\Delta, \tag{12}$$

$$k_1 = h(u(t) + \frac{k_0}{2}, t + \frac{\Delta}{2})\Delta, \tag{13}$$

$$k_2 = h(u(t) + \frac{k_1}{2}, t + \frac{\Delta}{2})\Delta, \tag{14}$$

$$k_3 = h(u(t) + k_2, t + \Delta)\Delta, \tag{15}$$

$$u(t + \Delta) = u(t) + \frac{1}{6}[k_0 + 2k_1 + 2k_2 + k_3]. \tag{16}$$

The function h is evaluated first at the left end-point of the interval, twice at extrapolated center points, and finally at the extrapolated right-endpoint. In Runge's solution, the interval $(t, t + \Delta)$ is divided into the four subintervals $\frac{\Delta}{6}, \frac{\Delta}{3}, \frac{\Delta}{3}$ and $\frac{\Delta}{6}$, which serve as the weights to be given to the four function evaluations.

The Runge-Kutta procedure and Euler's method share the property that only the value of the function u(t) at one point is necessary to start the procedure. Methods such as these are sometimes called "marching" or "open" methods. The major disadvantage of the Runge-Kutta methods is that they require several evaluations of the function h(u, t) per integration step. This increases the computing time necessary to produce a solution. The disadvantage of Euler's method is the rather large truncation error.

To combat the undesirable features of Euler's method and the Runge-Kutta procedure, a more sophisticated approach is necessary. The former methods assume that the slope, u'(t), is constant over the interval (t, t + Δ), a situation which is almost never the case in practice. It follows that a more accurate procedure would be to introduce a method whereby the slope is approximated by a set of polynomials and the polynomials used to <u>predict</u> the value u(t + Δ). Having predicted the value u(t + Δ), an iterative procedure utilizing the value u'(t + Δ) obtained from the original differential equation (1), is used to <u>correct</u> the predicted u(t + Δ). Methods of this type are called "predictor-corrector" methods and, as we shall see, often permit us to overcome the obstacles presented by Euler's method and the

Runge–Kutta procedures.

In order to obtain the predictor, let us pass an interpolating polynomial through the N known points $u'(t), u'(t - \Delta), \ldots, u'(t - (N - 1)\Delta)$ and then extrapolate the resulting function over the interval of integration. In order to start the integration, the first N points must be calculated via some other procedure, such as the Runge–Kutta method.

Using $u'(t)$ at a number of preceding points, we can employ for $u'(t^*)$ the interpolation formula

$$u'(t^*) = u'(t) + \frac{1}{\Delta} (t^* - t)\nabla u'(t)$$

$$+ \frac{1}{2!\Delta^2} (t^* - t)(t^* - (t - \Delta))\nabla^2 u'(t)$$

$$+ \frac{1}{3!\Delta^3} (t^* - t)(t^* - (t - \Delta))(t^* - (t - 2\Delta))\nabla^3 u'(t)$$

$$+\cdots, \quad t < t^* < t + \Delta, \tag{17}$$

where ∇ is the backwards difference operation $\nabla z(t) = z(t) - z(t - \Delta)$. Integrating (17) over the interval $(t, t + \Delta)$ gives the Adams–Bashforth formula

$$u(t + \Delta) = u(t) + \Delta[u'(t) + \frac{1}{2} \nabla u'(t) +$$

$$\frac{5}{12} \nabla^2 u'(t) + \frac{3}{8} \nabla^3 u'(t)] +\cdots. \tag{18}$$

Similarly, integrating (17) over the interval $(t - \Delta, t)$

yields the Adams-Moulton formula

$$u(t) = u(t - \Delta) + \Delta[u'(t) - \frac{1}{2} \nabla u'(t) -$$

$$\frac{1}{12} \nabla^2 u'(t) - \frac{1}{24} \nabla^3 u'(t) - \cdots], \quad (19)$$

or, upon replacing the differences by their definitions in

terms of $u'(t), u'(t - \Delta), \ldots,$ replacing t by $t + \Delta$ and

keeping only terms through the third order, we obtain,

$$u(t + \Delta) = u(t) + \frac{\Delta}{24} [55u'(t) - 59u'(t - \Delta) +$$

$$37u'(t - 2\Delta) - 9u'(t - 3\Delta)]. \quad (20)$$

The iterative corrector formula, utilizing $u'(t + \Delta)$,

can be derived by passing our interpolating polynomial through

the N points $u'(t + \Delta), u'(t), \ldots, u'(t - (N - 2)\Delta)$. Using

a Gregory-Newton backwards interpolation formula (see refer-

ences for details), the resulting expression is

$$u'(z) = u'(t + \Delta) + \nabla u'(t + \Delta)z + \frac{\nabla^2}{2!} u'(t + \Delta)z(z + \Delta)$$

$$+ \frac{\nabla^3}{3!} u'(t + \Delta)z(z + \Delta)(z + 2\Delta) + \cdots \frac{\nabla^N}{N!}$$

$$u'(t + \Delta)z(z + \Delta)\cdots(z + N\Delta - \Delta)$$

where

$$z = \frac{t^* - (t + \Delta)}{\Delta}, \qquad t \le t^* \le t + \Delta.$$

Integration of Eq. (1) with the substitution of $u'(z)$ for $u'(t)$ over the interval $(-\Delta, 0)$ in z, yields

$$u(t + \Delta) = u(t) + \Delta[u'(t + \Delta) - \frac{1}{2} \nabla u'(t + \Delta) -$$

$$\frac{1}{12} \nabla^2 u'(t + \Delta) - \cdots]. \quad (22)$$

Upon replacing the differences by their definition and retaining only terms through the third difference, we have

$$u(t + \Delta) = u(t) + \frac{\Delta}{24} [9u'(t + \Delta) + 19u'(t) -$$

$$5u'(t - \Delta) + u'(t - 2\Delta)]. \quad (23)$$

The complete predictor-corrector scheme used for all the numerical examples given throughout this book is the Adams-Moulton scheme just derived. Eq. (20) is used as the predictor, and Eq. (23) as the corrector. The computer program for this integration scheme will be found in the Appendix.

A local error estimate $E(\Delta)$ for the Adams-Moulton

integration scheme, assuming $u^{(5)}(t)$ does not change too

rapidly in the interval $(t - 3\Delta,\ t + \Delta)$, is given by

$$E(\Delta) = \frac{3}{8} \Delta^5 u^{(5)}(\xi) = 0(\Delta^5),\quad t - 3\Delta \le \xi \le t + \Delta.$$

(24)

Comparing the estimate (24) with that of the Runge-Kutta

procedure, we see that they are both $0(\Delta^5)$. However, upon

examination of (20) and (23), we see that the function

$h(u,\ t) = u'(t)$ must be evaluated only twice in the Adams-

Moulton scheme while the Runge-Kutta method required four

function evaluations per integration step. Thus, for the same

order of accuracy, the Adams-Moulton procedure will be

approximately twice as fast as the Runge-Kutta method at the

expense of a slight increase in computer storage required to

store the previous points.

Since a differential equation of second or higher

order may be reduced to a system of first-order equations

through the introduction of auxiliary dependent variables,

the above methods may be applied in the multidimensional as

well as the scalar case. For example, if a pair of simul-

taneous equations of the form

$$u' = h_1(u,\ v,\ t),\quad u(0) = c_1$$

$$v' = h_2(u,\ v,\ t),\quad v(0) = c_2$$

(25)

is to be integrated by the Runge-Kutta method described

above, the requisite formulas are

$$k_0 = h_1(u(t), v(t), t)\Delta, \tag{26}$$

$$n_0 = h_2(u(t), v(t), t)\Delta, \tag{27}$$

$$k_1 = h_1(u(t) + \frac{k_0}{2}, v(t) + \frac{n_0}{2}, t + \frac{\Delta}{2})\Delta, \tag{28}$$

$$n_1 = h_2(u(t) + \frac{k_0}{2}, v(t) + \frac{n_0}{2}, t + \frac{\Delta}{2})\Delta, \tag{29}$$

$$k_2 = h_1(u(t) + \frac{k_1}{2}, v(t) + \frac{n_1}{2}, t + \frac{\Delta}{2})\Delta, \tag{30}$$

$$n_2 = h_2(u(t) + \frac{k_1}{2}, v(t) + \frac{n_1}{2}, t + \frac{\Delta}{2})\Delta, \tag{31}$$

$$k_1 = h_1(u(t) + k_2, v(t) + n_2, t + \Delta)\Delta, \tag{32}$$

$$n_3 = h_2(u(t) + k_2, v(t) + n_2, t + \Delta)\Delta, \tag{33}$$

and

$$u(t + \Delta) = u(t) + \frac{1}{6}[k_0 + 2k_1 + 2k_2 + k_3], \tag{34}$$

$$v(t + \Delta) = v(t) + \frac{1}{6}[n_0 + 2n_1 + 2n_2 + n_3]. \tag{35}$$

Similar formulas hold for higher dimensional systems. In an

analogous fashion, the multidimensional counterparts of (20)

and (23) for the Adams-Moulton method can be developed.

5. STABILITY AND ERROR ANALYSIS

In this section, we briefly touch on one of the fundamental aspects of the numerical analysis and solution of
initial value problems--the stability of the method employed
and the minimization of some suitable measure of error. As
anyone who has even a nodding acquaintance with the numerical
solution of differential equations will immediately realize,
the concepts of stability of a method and measure of error
cannot be divorced from each other, the former essentially
being a subset of the latter. However, when undertaking the
error analysis of a particular problem, it is important to
recognize the various sources of potential error so, for
pedagogical purposes, we will artificially separate from
the overall error analysis certain "inseparable" ideas such
as stability, having forewarned the reader of the actual state
of affairs.

To introduce the concept of a stable numerical solution, consider the recursion relation

$$u_{i+1} = T_i(u_i, u_{i-1}, \ldots, u_1), \quad u_1 = c,$$

$$i = 1, 2, \ldots, N \tag{1}$$

where the sequence $\{u_i\}$ is a collection of real numbers and $\{T_i\}$ is a sequence of arithmetic transformations whose domain is the set of all sequences of real numbers. We shall call (1) a "numerical process." Clearly, all the integration formulas discussed in the previous section can be considered as special cases of (1). We will be interested in examining the numerical process for large N.

In practice, we cannot carry out the determination of the sequence $\{u_i\}$ as prescribed by (1) since, during the evaluation of each equation, we commit certain errors and proceed on the basis of an incorrect number to perform the next evaluation. Thus, we actually obtain the sequence $\{\bar{u}_i\}$ determined by

$$\bar{u}_{i+1} = T_i(\bar{u}_i, \bar{u}_{i-1}, \ldots, u_1) + \delta_i, \quad u_1 = c, \qquad (2)$$

where δ_i is usually a small number whose precise value is determined by such considerations as the values of $\bar{u}_k$, $k = 1, 2, \ldots, i$, round-off error, and so forth.

We are now faced with the question of whether or not δ_i stays sufficiently small for all i. Intuitively, we feel a solution is numerically stable if it doesn't "blow up too fast." More precisely, if given $\epsilon > 0$, there is a $\delta > 0$ such that $||u_1 - \bar{u}_1|| < \delta$ implies that

$||u_i - \bar{u}_i|| < \varepsilon$ for all i, then the solution $\{\bar{u}_i\}$ will be said to be stable. In this case, $||\cdot||$ denotes a suitable norm on the space of sequences containing $\{u_i\}$ and $\{\bar{u}_i\}$.

It is important to note that numerical stability is defined only in terms of the solution and not for the process itself. Since the solution depends only upon the starting value u_1, a different starting value may change the entire behavior of the solution depending upon the operators T_i. Thus, it is meaningful to talk only of numerical stability for a single solution.

As mentioned above, the main factors affecting the numerical stability of a solution are the structure of the computer program, the number of arithmetical operations called for by the operators T_i, and computer word size (round-off error). All of these factors should be considered before dispatching a particular problem to the computer.

Besides stability, the prime factors affecting the overall accuracy of the above mentioned integration methods are the behavior of the function $h(u, t)$, the interval length under consideration, and the integration step size. For the majority of cases, we have no control over the function h or the interval length as they are generally dictated by the particular problem at hand. However, the integration step size may be varied to maintain some measure of control on the accuracy. For example, in the case of the solution of an

equation by Euler's method, it can be shown that an indepen-

dent error estimate ε_n at the point t_n is given by

$$|\varepsilon_n| \leq \frac{\Delta N_1}{2K} (\exp(k(t_n - t_0)) - 1) \tag{3}$$

where t_0 is the initial point of the interval, $t_n = t_0 +$

$n\Delta$, K is the Lipshitz constant for the function h, and N_1

is a bound on h(u, t). The importance of (3) is that it

explicitly demonstrates that, at a fixed point in the interval

(t_0, t_N), the error tends to zero as $\Delta \to 0$. Similar error

estimates for the Runge-Katta and Adams-Moulton procedures,

although they are not as simple as (3), show that in most

cases one can expect the solution accuracy to increase as Δ

decreases until round-off error becomes the dominant factor.

At the point where round-off error enters as a serious con-

tributor to overall inaccuracy, the numerical analyst finds

himself precariously perched upon the horns of a dilemma.

Further decrease in the step size Δ will only increase

round-off due to the fixed number of significant digits that

the computer can maintain at each step; however, with respect

to truncation error, a decrease in Δ is desirable as indi-

cated by (3). It is this gray area that computing becomes

more of an art than a science and past experience becomes the

guide. As a practical matter, one simple rule to follow is
to solve the problem using one step size, then halve the
original step size and solve again. If the two answers are
in sufficient agreement, the solution is accepted; if not,
further experimentation is necessary. This procedure allows
us to check the self-consistency of a numerical integration
scheme and, through appropriate combination of the resulting
solutions, often enables a solution to be constructed which
is more accurate than either of the two component solutions.
References to this technique are given at the end of the
chapter.

6.   INITIAL VALUE VS. TWO-POINT BOUNDARY VALUE PROBLEMS

Our discussion so far has concerned itself solely
with initial value problems where all the conditions necessary to determine a solution are given at a single point.
However, in many problems in mathematical physics, engineering, economics, and biology conditions are specified at two
or more points and, consequently, fail to fall within the
class of problems treated above.

For example, consider the scalar two-point boundary
value problem

$$u'' + k_1 u = 0 \tag{1}$$

$$u(0) = c_1, \quad u(T) = c_2, \qquad 0 \le t \le T, \tag{2}$$

and k_1, c_1 and c_2 are constants. The general solution
of (1) is

$$u = a_1 \exp(\alpha_1 t) + a_2 \exp(\alpha_2 t), \; \alpha_1 \ne \alpha_2. \tag{3}$$

Using the boundary conditions (2), we obtain the set of
linear algebraic equations for a_1 and a_2 as

$$a_1 + a_2 = c_1, \; \exp(\alpha_1 T) a_1 + \exp(\alpha_2 T) a_2 = c_2. \tag{4}$$

From the theory of equations, we know that (4) possesses a
unique solution if the determinant

$$D(T) = \begin{vmatrix} 1 & 1 \\ \exp(\alpha_1 T) & \exp(\alpha_2 T) \end{vmatrix} =$$

$$\exp(\alpha_2 T) - \exp(\alpha_1 T) \ne 0. \tag{5}$$

Equation (5) will certainly hold for all $T > 0$ if α_1 and
α_2 are real and distinct. However, if α_1 and α_2 are a
complex pair

$$\alpha_1 = x + iy,$$

$$\alpha_2 = x - iy,$$

then it is easy to see that

$$D(T) = \exp(\alpha_1 T) - \exp(\alpha_2 T) = e^{xT}(e^{-iyT} - e^{iyT}),$$

and, consequently, that the determinant is zero for $T = n\pi/y$, $n = 1,2,\ldots$.

The important point to note is that the two-point boundary condition is a vastly different matter than a one point condition as far as existence and uniqueness of solutions is concerned. Depending upon the interval length, it is easy to construct two-point boundary value problems which have one, none or an infinite number of solutions. This is a sharp contrast from the situation put forth in our existence and uniqueness theorem for initial value problems.

In addition to the formidable analytic difficulties of two-point boundary value problems, the computational aspects serve as a further inducement for the development of the methods we shall present in the remainder of the book. As shown above, the digital computer is ideally suited to carry out the iterative process of solving an initial value problem with a given set of initial conditions. In the case of a two-point boundary condition, we do not have a sufficient number of conditions at either point to apply the standard integration procedures described above. Consequently, we must often resort to a number of strategems, tricks, and

"seat of the pants" guesswork to succsssfully produce a
numerical solution. In this era of fully automated procedures, this is a confession of weakness and lack of knowledge concerning the underlying structure of the problem
considered. In the remaining chapters, we will treat many
of the types of functional equations occurring in physics and
engineering, constantly keeping in mind the goal of reformu-
lation as an initial value problem.

CHAPTER TWO

NOTES AND REFERENCES

§2.   Classic references to the treatment of matrices and

differential equations are the books

Bellman, R., Introduction to Matrix Analysis,

McGraw-Hill, New York, 1969, Chapter 10,

Gantmacher, F. R., The Theory of Matrices, Chelsea

Pub. Co., New York, 1960, Chapter 14.

Extensive application of vector-matrix theory to

numerous problems of applied mathematics, physics,

and engineering have been made by many authors. See:

Lichnerowicz, A., Linear Algebra and Analysis,

Holden-Day Co., San Francisco, 1967,

Nering, E., Linear Algebra and Matrix Theory,

John Wiley, New York, 1963,

Hohn, F., Elementary Matrix Algebra, Macmillan

Publishing Co., New York, 1964.

§3.   An introductory account of the material given here

may be found in:

Bellman, R., Modern Elementary Differential

Equations, Addison-Wesley Co., Reading, Mass., 1967.

More advanced results for ordinary differential equations are given in:

Coddington, E. and N. Levinson, Theory of Ordinary Differential Equations, McGraw-Hill, New York, 1955,

Ince, E., Ordinary Differential Equations, Dover, New York, 1944,

Hartman, P., Ordinary Differential Equations, Wiley, New York, 1964.

§4.   A detailed treatment of this material may be found in almost any textbook on numerical analysis. Particularly recommended are:

Collatz, L., The Numerical Treatment of Differential Equations, Springer-Verlag, New York, 1966,

Babuska, I., M. Prager, and E. Vitasek, Numerical Processes in Differential Equations, Interscience, New York, 1966,

Isaacson, E., and H. Keller, Analysis of Numerical Methods, Wiley, New York, 1966,

Henrici, P., Discrete Variable Methods in Ordinary Differential Equations, Wiley, New York, 1962.

§5.   The analytic (as contrasted with the numerical) aspects of stability theory are extensively covered in:

Bellman, R., Stability Theory of Differential Equations, McGraw-Hill, New York, 1953,

LaSalle, J. and S. Lefschetz, Stability by Liapunov's Direct Method with Applications, Academic Press, New York, 1961,

Cesari, L., Asymptotic Behavior and Stability Problems in Ordinary Differential Equations, Springer-Verlag, New York, 1963.

The study of the numerical stability of various finite difference algorithms is treated in

Henrici, P., Error Propagation for Difference Methods, Wiley, New York, 1963,

Dahlquist, G., "Stability and Error Bounds in the Numerical Integration of Ordinary Differential Equations," Trans. Roy. Inst. Tech., Stockholm, No. 130 (1959),

Wendroff, B., Theoretical Numerical Analysis, Academic Press, New York, 1966.

§6.   The numerical treatment of many linear and nonlinear

        two-point boundary value problems is taken up in:

        Bellman, R. and R. Kalaba, Quasilinearization and

        Nonlinear Boundary Value Problems, Elsevier, New

        York, 1965,

        Keller, H., Numerical Methods for Two-Point

        Boundary Value Problems, Ginn-Blaisdell, Waltham,

        Mass., 1968,

        Lee, S., Quasilinearization and Invariant Imbedding,

        Academic Press, New York, 1968.

CHAPTER THREE

TWO-POINT BOUNDARY VALUE PROBLEMS

1. INTRODUCTION

The objects of our attention in this chapter, two-point boundary value problems, constitute perhaps the simplest member of the hierarchy of problems to which our methods may be applied. This is not to imply that there is anything particularly simple or routine about the mathematics of two-point boundary value problems, since, by any standards, the theory of such problems is far from complete and is still being vigorously pursued. But rather, our claim is that two-point boundary value problems possess the simplest nontrivial structure exhibiting the unpleasant features which our theory has been developed to help remedy. As a result, the ideas presented in this chapter serve as a guide for the treatment of more complex functional equations.

During the course of this chapter, we shall examine several aspects of both linear and nonlinear two-point boundary value problems from an initial value point of view. We begin by treating simple linear systems, illustrating the techniques and results by an example involving Poisson's equation. A discussion of the application of our ideas for the construction of Green's functions is also given. Following these results for linear systems, we next consider various types of nonlinear problems. A derivation of the initial value system is presented, together with a validation, showing that the initial value system is indeed mathematically equivalent to the boundary value problem.

2.   LINEAR TWO-POINT BOUNDARY VALUE PROBLEMS

We begin our study of two-point boundary value problems by considering the scalar equations

$$\dot{x}(t) = a(t)x(t) + b(t)y(t),$$

$$\alpha_1 x(0) + \alpha_2 y(0) = 0, \tag{1}$$

$$\dot{y}(t) = c(t)x(t) + d(t)y(t) + f(t),$$

$$\alpha_3 x(T) + \alpha_4 y(T) = 1, \tag{2}$$

where a, b, c, d, f are continuous on $0 \le t \le T$, and

$\{\alpha_i\}$ are constants, i = 1,2,3,4. Although (1)-(2) is not

the most general form of a linear two-point boundary value

problem, it suffices to cover most problems of interest in

physics and engineering. The interested reader will find no

difficulty in extending the ideas that follow to cover more

general problems involving multidimensional versions of (1)-

(2) as well as somewhat different boundary conditions. We

shall give some results along these lines in a later section.

Since a complete set of conditions uniquely specify-

ing x and y are not given at a single point but rather at

the two points t = 0 and t = T, (1)-(2) is termed a two-

point boundary value problem. This two-point condition

greatly complicates both the analysis and computation of the

solution to (1)-(2) since not enough information is prescribed

at any one point to completely specify the vector field

(x(t), y(t)). From a computational standpoint, this corres-

ponds to a situation in which straightforward use of various

numerical integration schemes such as Runge-Kutta, Adams-

Bashforth, etc., is precluded due to a lack of sufficient

information at a point to get the algorithm started. Need-

less to say, these difficulties become even more acute for

nonlinear problems.

Our goal is to derive an initial value, rather than

two-point boundary value, problem which "represents" the

solution to (1)-(2). By "represents", we mean that the solution of the initial value (or Cauchy) system uniquely determines the solution to the boundary value problem, and conversely. In mathematical parlance, the two problems are isomorphic and, as a result, properties established concerning one system are immediately transferable to the other. Since our aims are primarily pedagogical, most of our results will be formally derived laying aside rigorous details (which, in some cases, are highly nontrivial). Let us now turn our attention to the task at hand.

The imbedding parameter we shall use to derive a Cauchy system is the interval length T. Consequently, we rewrite (1)-(2) to explicitly indicate the dependence of the solution upon T:

$$\dot{x}(t, \ T) = a(t)x(t, \ T) + b(t)y(t, \ T),$$
$$\alpha_1 x(0, \ T) + \alpha_2 y(0, \ T) = 0, \tag{3}$$

$$\dot{y}(t, \ T) = c(t)x(t, \ T) + d(t)y(t, \ T) + f(t),$$
$$\alpha_3 x(T, \ T) + \alpha_4 y(T, \ T) = 1, \qquad 0 \le t \le T. \tag{4}$$

To reduce the problem to more digestable proportions, we make use of linearity and consider the two systems:

<u>System I</u>

$$\dot{u}(t, T) = a(t)u + b(t)v,$$

$$\alpha_1 u(0, T) + \alpha_2 v(0, T) = 0, \qquad\qquad (5)$$

$$\dot{v}(t, T) = c(t)u + d(t)v + f(t), \quad \leftarrow with\ input$$

$$\alpha_3 u(T, T) + \alpha_4 v(T, T) = 0, \qquad 0 \le t \le T, \qquad (6)$$

and

<u>System II</u>

$$\dot{p}(t, T) = a(t)p + b(t)q,$$

$$\alpha_1 p(0, T) + \alpha_2 q(0, T) = 0, \qquad\qquad (7)$$

$$\dot{q}(t, T) = c(t)p + d(t)q, \quad \leftarrow no\ input$$

$$\alpha_3 p(T, T) + \alpha_4 q(T, T) = 1. \qquad\qquad (8)$$

The superposition principle for linear systems then allows us
to write

$$x(t, T) = u(t, T) + p(t, T), \qquad\qquad (9)$$

$$y(t, T) = v(t, T) + q(t, T), \qquad 0 \le t \le T. \qquad (10)$$

Henceforth, we shall concern ourselves with systems I and II.

Let us first consider the functions u and v of
system I. We wish to examine how the solution curves change
at a fixed point t, $0 \le t \le T$, as the interval length T is

changed. Differentiating (5)-(6) with respect to T gives

$$\dot{u}_T(t, T) = a(t)u_T(t, T) + b(t)v_T(t, T),$$

$$\alpha_1 u_T(0, T) + \alpha_2 v_T(0, T) = 0, \qquad\qquad (11)$$

$$\dot{v}_T(t, T) = c(t)u_T(t, T) + d(t)v_T(t, T),$$

$$\alpha_3[\dot{u}(T, T) + u_T(T, T)] +$$

$$\alpha_4[\dot{v}(T, T) + v_T(T, T)] = 0, \qquad 0 \le t \le T. \qquad (12)$$

Here a dot represents differentiation with respect to t,
$( )_T$ differentiation with respect to T. Comparing (11)-
(12) with (7)-(8), we see that

$$u_T(t, T) = -[\alpha_3\dot{u}(T, T) + \alpha_4\dot{v}(T, T)]p(t, T), \qquad (13)$$

$$v_T(t, T) = -[\alpha_3\dot{u}(T, T) + \alpha_4\dot{v}(T, T)]q(t, T),$$

$$0 \le t \le T. \qquad\qquad (14)$$

We now consider the bracketed term in Eqs. (13)-(14).
From Eqs. (5)-(6) with t = T, we have

$$\dot{u}(T, T) = a(T)u(T, T) + b(T)v(T, T), \qquad (15)$$

$$\dot{v}(T, T) = c(T)u(T, T) + d(T)v(T, T) + f(T). \qquad (16)$$

Introduce the new variables m and n by

$$m(T) = u(T, T),$$ (17)

$$n(T) = v(T, T), \qquad T \geq 0.$$ (18)

In view of (15)-(16), it suffices to determine the functions m and n. We first deal with m.

Differentiate Eq. (13) to obtain

$$m'(T) = \dot{u}(T, T) + u_T(T, T) = a(T)m(T) +$$
$$b(T)n(T) - \{\alpha_3[a(T)m(T) + b(T)n(T)] +$$
$$\alpha_4[c(T)m(T) + d(T)n(T) + f(T)]\}p(T, T),$$
$$T > 0.$$ (19)

Similarly, for n we have

$$n'(T) = \dot{v}(T, T) + v_T(T, T) = c(T)m(T) +$$
$$d(T)n(T) + f(T) - \{\alpha_3[a(T)m(T) + b(T)n(T)] +$$
$$\alpha_4[c(T)m(T) + d(T)n(T) + f(T)]\}q(T, T),$$
$$T > 0.$$ (20)

Eqs. (19) and (20) show that we must consider the quantities $p(T, T)$ and $q(T, T)$.

Differentiate Eqs. (7) and (8) with respect to T. This yields

$$\dot{p}_T(t, T) = a(t)p_T + b(T)q_T,$$
$$\alpha_1 p_T(0, T) + \alpha_2 q_T(0, T) = 0,$$ (21)

$$\dot{q}_T(t, T) = c(t)p_T + d(t)q_T,$$

$$\alpha_3[\dot{p}(T, T) + p_T(T, T)] + \alpha_4[\dot{q}(T, T) +$$

$$q_T(T, T)] = 0, \qquad 0 \le t \le T. \tag{22}$$

Comparing (7)-(8) with (21)-(22) shows that

$$p_T(t, T) = -[\alpha_3\dot{p}(T, T) + \alpha_4\dot{q}(T, T)]p(t, T) \tag{23}$$

and

$$q_T(t, T) = -[\alpha_3\dot{p}(T, T) + \alpha_4\dot{q}(T, T)]q(t, T),$$

$$0 \le t \le T. \tag{24}$$

To make use of these relations, we observe that from (7) and (8) with $t = T$

$$\dot{p}(T, T) = a(T)p(T, T) + b(T)q(T, T), \tag{25}$$

$$\dot{q}(T, T) = c(T)p(T, T) + d(T)q(T, T). \tag{26}$$

Let the functions r and s be given by

$$r(T) = P(T, T), \tag{27}$$

$$s(T) = q(T, T). \tag{28}$$

We now derive a Cauchy system satisfied by r and s. First differentiate r to obtain

$$r'(T) = \dot{p}(T, T) + p_T(T, T) = a(T)r(T) +$$

$$b(T)s(T) - \{\alpha_3[a(T)r + b(T)s] +$$

$$\alpha_4[c(T)r + d(T)s]\}r. \tag{29}$$

In a similar manner, the equation for s is seen to be

$$s'(T) = c(T)r + d(T)s - \{\alpha_3[a(T)r + b(T)s] +$$

$$\alpha_4[c(T)r + d(T)s]\}s, \qquad T > 0. \tag{30}$$

Combining terms in (29)-(30) gives

$$r'(T) = b(T)s + r[a(T) - \alpha_3 b(T)s -$$

$$\alpha_4 d(T)s] - r^2[\alpha_3 a(T) + \alpha_4 c(T)], \tag{31}$$

$$s'(T) = c(T)r + s[d(T) - \alpha_3 a(T)r -$$

$$\alpha_4 c(T)r] - s^2[\alpha_3 b(T) + \alpha_4 d(T)], \qquad T > 0. \tag{32}$$

The initial conditions at $T = 0$ are obtained from Eqs. (7) and (8) by solving the system

$$\alpha_1 r(0) + \alpha_2 s(0) = 0, \tag{33}$$

$$\alpha_3 r(0) + \alpha_4 s(0) = 1. \tag{34}$$

Obviously, this forces the compatability condition

$\alpha_1\alpha_4 - \alpha_2\alpha_3 \neq 0$ to insure a unique solution exists.

Knowledge of r and s allows us to obtain p and q from the initial value system (23) and (24). The equations are

$$p_T(t, T) = - \{r[\alpha_3 a(T) + \alpha_4 c(T)] +$$

$$s[\alpha_3 b(T) + \alpha_4 d(T)]\}p(t, T), \tag{35}$$

$$q_T(t, T) = - \{r[\alpha_3 a(T) + \alpha_4 c(T)] +$$

$$s[\alpha_3 b(T) + \alpha_4 d(T)]\}q(t, T), \qquad 0 \leq t \leq T. \tag{36}$$

The initial conditions at $T = t$ are

$$p(t, t) = r(t), \tag{37}$$

$$q(t, t) = s(t). \tag{38}$$

Returning now to Eqs. (19) and (20) for the functions m and n, we see that

$$m'(T) = a(T)m + b(T)n - \{m[\alpha_3 a(T) +$$

$$\alpha_4 c(T)] + n[\alpha_3 b(T) + \alpha_4 d(T)] + f(T)\}r(T), \tag{39}$$

$$n'(T) = c(T)m + d(T)n + f(T) - \{m[\alpha_3 a(T) +$$

$$\alpha_4 c(T)] + n[\alpha_3 b(T) + \alpha_4 d(T)] + f(T)\}s(T),$$

$$T > 0. \tag{40}$$

The initial conditions at $T = 0$ are given by Eqs. (5) and (6) as

$$m(0) = 0, \tag{41}$$

$$n(0) = 0. \tag{42}$$

The equations for u and v are determined in terms of m and n by

$$u_T(t,\ T) = - \{m[\alpha_3 a(T) + \alpha_4 c(T)] +$$
$$n[\alpha_3 b(T) + \alpha_4 d(T)] + f(T)\}p(t,\ T), \tag{43}$$

$$v_T(t,\ T) = -\{m[\alpha_3 a(T) + \alpha_4 c(T)] +$$
$$n[\alpha_3 b(T) + \alpha_4 d(T)] + f(T)\}q(t,\ T), \tag{44}$$

$$0 \le t \le T.$$

At $T = t$, we have

$$u(t,\ t) = m(t), \tag{45}$$

$$v(t,\ t) = n(t). \tag{46}$$

This completes our derivation of the complete Cauchy system for determining the functions u, v, p, and q necessary to obtain x and y, the solutions to (1)-(2).

3.   SUMMARY OF THE CAUCHY SYSTEM

Since the derivation given in the last section was somewhat lengthy, we now summarize and collect in one place the relevant equations.   The complete Cauchy system consists of the following equations for the functions   r, s, m, n, p, q, u, and   v:

$$r'(T) = b(T)s + r[a(T) - \alpha_3 b(T)s - \alpha_4 d(T)s]$$
$$- r^2[\alpha_3 a(T) + \alpha_4 c(T)], \tag{1}$$

$$s'(T) = c(T)r + s[d(T) - \alpha_3 a(T)r - \alpha_4 c(T)r]$$
$$- s^2[\alpha_3 b(T) + \alpha_4 d(T)], \tag{2}$$

$$m'(T) = a(T)m + b(T)n - \{m[\alpha_3 a(T) +$$
$$\alpha_4 c(T)] + n[\alpha_3 b(T) + \alpha_4 d(T)] + f(T)\}r(T), \tag{3}$$

$$n'(T) = c(T)m + d(T)n + f(T) - \{m[\alpha_3 a(T) +$$
$$\alpha_4 c(T)] + n[\alpha_3 b(T) + \alpha_4 d(T)] + f(T)\}s(T), \tag{4}$$

$$p_T(t, T) = -\{r[\alpha_3 a(T) + \alpha_4 c(T)] +$$
$$s[\alpha_3 b(T) + \alpha_4 d(T)]\}p(t, T), \tag{5}$$

$$q_T(t, T) = -\{r[\alpha_3 a(T) + \alpha_4 c(T)] +$$
$$s[\alpha_3 b(T) + \alpha_4 d(T)]\}q(t, T), \tag{6}$$

$$u_T(t, T) = - \{m[\alpha_3 a(T) + \alpha_4 c(T)] +$$

$$n[\alpha_3 b(T) + \alpha_4 d(T)] + f(T)\}p(t, T), \quad (7)$$

$$v_T(t, T) = - \{m[\alpha_3 a(T) + \alpha_4 c(T)] +$$

$$n[\alpha_3 b(T) + \alpha_4 d(T)] + f(T)\}q(t, T), \quad (8)$$

$$0 \le t \le T.$$

The initial conditions are given by

$$\alpha_1 r(0) + \alpha_2 s(0) = 0$$

$$\alpha_3 r(0) + \alpha_4 s(0) = 1, \quad (9)$$

$$m(0) = 0, \quad (10)$$

$$n(0) = 0, \quad (11)$$

$$p(t, t) = r(t), \quad (12)$$

$$q(t, t) = s(t), \quad (13)$$

$$u(t, t) = m(t), \quad (14)$$

$$v(t, t) = n(t). \quad (15)$$

The solution curves for the original system are then given by

$$x(t, T) = u(t, T) + p(t, T), \quad (16)$$

$$y(t, T) = v(t, T) + q(t, T).\tag{17}$$

Suppose the solution is desired at a set of abscissas $0 \leq t_1 < t_2 < t_3 < \cdots < t_N \leq T^*$, where T^* is the interval length of interest. The solution procedure is to integrate the equations for r, s, m, and n from $T = 0$ to $T = t_1$. At this point equations for the functions $p(t_1, T)$, $q(t_1, T)$, $u(t_1, T)$, and $v(t_1, T)$ are adjoined with the initial conditions given by (12)-(15). The entire system is then integrated from $T = t_1$ to $T = t_2$, at which point additional equations for the functions $p(t_2, T)$, $q(t_2, T)$, $u(t_2, T)$, and $v(t_2, T)$ are adjoined with the appropriate initial conditions. This procedure is carried out for each t_i and the integration continues until $T = T^*$. At this point, the desired solution values are obtained from the functions $p(t_i, T^*)$, $q(t_i, T^*)$, $u(t_i, T^*)$, and $v(t_i, T^*)$ by means of the relations (16) and (17). Notice that as the integration proceeds, solution values may be obtained for all interval lengths less than T^*. This feature provides a parameter study which may be of value in many situations.

4. AN UNSTABLE EXAMPLE

To illustrate one facet of our theory, consider the simple linear system

$$\dot{x} = 10y, \quad x(0) = 0, \tag{1}$$

$$\dot{y} = 10x, \quad y(2) = 1. \tag{2}$$

The characteristic values of this system are ± 10. Since the interval length of interest $(T = 2)$ is not small, producing two accurate independent solutions on a computer by a marching method is not a simple task.

The relevant quantities for our initial value system are $\alpha_1 = \alpha_4 = 1$, $\alpha_2 = \alpha_3 = 0$, $a(t) = d(t) = f(t) = 0$, $b(t) = c(t) = 10$. Notice that since $f \equiv 0$, only the functions p and q are of interest. The Cauchy system then becomes

$$r'(T) = 10 - 10r^2, \quad r(0) = 0, \tag{3}$$

$$p_T(t, T) = -10r(T)p(t, T), \quad p(t, t) = r(t), \tag{4}$$

$$q_T(t, T) = -10r(T)q(t, T), \quad q(t, t) = 1. \tag{5}$$

The nature of the boundary conditions obviously implies $s(T) \equiv 1$. From Eq. (3) we see that $0 \leq r(T) \leq 1$, $T \geq 0$. In view of this bound, Eqs. (4) and (5) have no increasing solutions as T increases. For this simple example, this can also be seen from the closed form solutions

$$r(T) = \tanh 10T, \quad T \geq 0, \tag{6}$$

$$p(t, T) = \frac{\sinh 10t}{\cosh 10T} = x(t, T), \tag{7}$$

$$q(t, T) = \frac{\cosh 10t}{\cosh 10T} = y(t, T), \qquad 0 \le t \le T. \tag{8}$$

The foregoing problem points out a feature of our methods that will frequently reappear in later sections, namely, that although the original boundary value problem may be numerically very unstable, the equivalent Cauchy system usually possesses a computationally stable solution. In more general situations where explicit solutions are not available, the stability is not obvious but may still be established under suitable hypotheses on the coefficient functions a, b, c, d. Many results of this type are found in the references cited at the end of the chapter.

5. MULTIDIMENSIONAL SYSTEMS

Oftentimes linear two-point boundary value problems occur as a system of equations having the form

$$\dot{x}(t) = A(t)x + B(t)y,$$
$$(x(0), \alpha_1) + (y(0), \alpha_2) = 0, \tag{1}$$

$$\dot{y}(t) = C(t)x + D(t)y + f(t),$$
$$(x(T), \alpha_3) + (y(T), \alpha_4) = 1, \qquad t \ge 0, \tag{2}$$

where x, y, f, α_1, α_2, α_3, α_4 are n-dimensional vectors, A,

B, C, D are n × n matrices, and (,) denotes the

usual inner product. The reader will find no difficulty in

obtaining an initial value representation for the solution of

this system by following the same steps given above, keeping

in mind the usual operations with vectors and matrices. It

will be seen that the corresponding Cauchy system will now

contain matrix Riccati equations possessing the same stability

properties indicated above for the scalar case.

6. POISSON'S EQUATION AND NUMERICAL INSTABILITY

To illustrate both the stability features and the

numerical effectiveness of the imbedding approach, let us con-

sider the solution of Poisson's equation in a rectangle by

means of Kantorovich's extension of the Rayleigh-Ritz proce-

dure. In addition, this example will also serve to illustrate

the treatment of a problem whose boundary conditions are not

covered by our previous development.

The problem of interest is

$$u_{xx} + u_{yy} = -2, \quad (x, y) \; \varepsilon \; R,$$

$$u(x, y) = 0, \qquad (x, y) \; \varepsilon \; \partial R, \qquad \qquad (1)$$

where R is the region of the x-y plane $\{(x, y): 0 \leq x \leq 1,$

$-1 \leq y \leq 1$}.   Physically, this problem arises in the deter-

mination of the equilibrium position of a membrane fixed at

the edges, displaced by a force of 2 units/unit area in the

positive z-direction. By standard separation of variables

arguments, it is easily seen that the solution of (1) is

given by the series expansion

$$u(x, y) = x(1 - x) - \frac{8}{\pi^3} \sum_{n=1,3,5\ldots} \frac{\cosh n\pi y \sin n\pi x}{n^3 \cosh n\pi}.$$

$$(2)$$

This form will be useful for purposes of checking the accu-

racy of the numerical procedure to be described below.

The solution to (1) is also determined by the minimum

of the quadratic functional

$$J = \int\int_R [w_x^2 + w_y^2 - 4w] \, dy \, dx, \qquad (3)$$

subject to the boundary conditions of (1). Classical varia-

tional theory (Dirichlet's principle) shows that the function

w minimizing J, subject to the appropriate boundary condi-

tions, provides the unique solution to (1).

Let us use a modification of the Rayleigh-Ritz

procedure which is due to Kantorovich to minimize J. We

seek the minimizing function   w   in the form

$$w(x, y) = f(x)(y^2 - 1).$$ (4)

Substituting (4) into (3), performing the indicated differentiations and the y integration, we obtain the variational
integral for the unknown function   f:

$$J(f) = \int_0^1 \left[\frac{16}{15} f'^2 + \frac{8}{3} f^2 + \frac{16}{3} f \right] dx.$$ (5)

The Euler equation associated with (5) is

$$f'' - \frac{5}{2} f = \frac{5}{2}.$$ (6)

For the function   w   to satisfy the boundary conditions of
(1), we must have

$$f(0) = f(1) = 0.$$ (7)

In this case, the two-point boundary value problem possesses
the explicit solution

$$f(x) = -1 + \cosh(\sqrt{10}\, x/2)$$

$$+ \left(\frac{1 - \cosh(\sqrt{10}/2)}{\sinh(\sqrt{10}/2)} \right) \sinh(\sqrt{10}\, x/2). \quad (8)$$

Even in this simple case the appearance of hyperbolic sines

and cosines indicate a possible computational instability

which may block attempts at a solution by ordinary numerical

methods. Consequently, let us use imbedding arguments to

derive an initial value problem to produce the function f.

Notice that the boundary conditions (7) preclude use of our

previous results since $\alpha_1\alpha_4 - \alpha_3\alpha_2 = 0.$

7. ANOTHER INITIAL VALUE SYSTEM

Let us regard the two-point boundary value problem for

f   as a special case of the slightly more general system

$$\dot{u}(t, T) = v(t, T), \quad u(0, T) = 0, \tag{1}$$

$$\dot{v}(t, T) = p(t) - q(t)v(t, T), \quad u(T, T) = 0,$$
$$0 \leq t \leq T. \tag{2}$$

Again, we explicitly denote the dependence of the functions

u   and   v   upon   T.   In conjunction with (1)–(2), we must

also consider the system

$$\dot{w}(t, T) = z(t, T), \quad w(0, T) = 0 \tag{3}$$

$$\dot{z}(t, T) = -q(t)w(t, T), \quad w(T, T) = 1,$$
$$0 \leq t \leq T. \tag{4}$$

To begin, differentiate (1) and (2) with respect to   T.   This yields

$$\dot{u}_T(t,\ T) = v_T(t,\ T),\quad u_T(0,\ T) = 0, \tag{5}$$

$$\dot{v}_T(t,\ T) = -q(t)v_T(t,\ T),$$

$$\dot{u}(T,\ T) + u_T(T,\ T) = 0,\qquad 0 \le t \le T. \tag{6}$$

Comparing (5)–(6) with (3)–(4), we see that

$$u_T(t,\ T) = -\dot{u}(T,\ T)w(t,\ T), \tag{7}$$

$$v_T(t,\ T) = -\dot{u}(T,\ T)z(t,\ T),\qquad 0 \le t \le T. \tag{8}$$

A similar differentiation of (3)–(4) gives the relation

$$w_T(t,\ T) = -\dot{w}(T,\ T)w(t,\ T), \tag{9}$$

$$z_T(t,\ T) = -\dot{w}(T,\ T)z(t,\ T),\qquad 0 \le t \le T. \tag{10}$$

We must now focus on the two quantities $\dot{w}(T,\ T)$ and $\dot{u}(T,\ T)$.

Introduce the new function $r(T)$ as

$$r(T) = \dot{w}(T,\ T) = z(T,\ T). \tag{11}$$

Differentiating r with respect to T and making use of Eqs. (4) and (10), we have

$$r'(T) = \dot{z}(T, T) + z_T(T, T)$$

$$= -q(T)w(T, T) - \dot{w}(T, T)z(T, T)$$

$$= -q(T) - r^2(T), \qquad T > 0. \tag{12}$$

This is a simple Riccati equation for r. For the moment, we shall set aside consideration of the initial condition at $T = 0$. Knowledge of the function r completes the determination of w and z through Eqs. (9) and (10), since the obvious initial conditions at $T = t$ from Eqs. (4) and (11) are

$$w(t, t) = 1, \tag{13}$$

$$z(t, t) = r(t). \tag{14}$$

Now let us consider the function $\dot{u}(T, T)$. Introduce $s(T)$ as

$$s(T) = \dot{u}(T, T) = v(T, T), \qquad T \geq 0. \tag{15}$$

Differentiating s with respect to T and using Eqs. (2) and (8), we have

$$s'(T) = \dot{v}(T, T) + v_T(T, T)$$

$$= p(T) - q(T)s(T) - s(T)r(T), \qquad T > 0. \tag{16}$$

The initial condition at $T = 0$ is seen from Eqs. (1)-(2)

(by continuity) to be

$$s(0) = 0. \tag{17}$$

Knowledge of s allows the determination of u and v

through Eqs. (7)-(8) since appropriate initial conditions

at T = t are

$$u(t, t) = 0, \tag{18}$$

$$v(t, t) = s(t). \tag{19}$$

The only item needed to completely describe our

Cauchy system is the initial condition at T = 0 for the

function r. Since r is defined to be $\dot{w}(T, T)$, from Eqs.

(3)-(4) it is seen that for T "small", $r(T) \approx \frac{1}{T}$. This

suggests expanding r about T = 0 in the Laurent series

$$r(T) = \frac{1}{T} + a_0 + a_1 T + a_2 T^2 + \cdots \tag{20}$$

valid for $0 < |T| \leq \varepsilon \ll 1$. Assuming that q is an analytic

function in some neighborhood about T = 0, we may expand in

the Taylor series

$$q(T) = q_0 + q_1 T + q_2 T^2 + \cdots. \tag{21}$$

Substituting the expansions (20) and (21) into Eq. (12) and

equating like powers of T, we may obtain the sequence $\{a_n\}$. The first few terms are $a_0 = 0$, $a_1 = -q_0/3$, $a_2 = -q_1/4$, $a_3 = -(q_2 + q_0^2/9)/5$, $a_4 = -(q_3 + q_0q_1/6)/6$, $a_5 = -(q_4 + 2q_0q_2/15 + 2q_0^3/135 + q_1^2/16)/7$. Thus, we use (20) to calculate the initial condition on r for some "small" value of T near 0.

8. NUMERICAL RESULTS

The equation we wish to solve for f (Eq. (6.7)) corresponds to the special case $q = -5/2$, $p = 5/2$, which leads to the series expansion

$$r(T) \cong \frac{1}{T} + \frac{5}{6} T - \frac{5}{36} T^3 + \frac{25}{756} T^5 + \cdots. \tag{1}$$

Fortunately, in this case we also have the luxury of the closed form solution

$$r(T) = \sqrt{5/2} \left[\frac{\exp(2\sqrt{5/2}\ T) + 1}{\exp(2\sqrt{5/2}\ T) - 1} \right]. \tag{2}$$

For various values of T, we obtain the results of Table 1.

Table 1:

Initial Conditions on r

T	Truncated Series (1)	Exact
0.1	10.08319	10.08319
0.2	5.16557	5.16557
0.3	3.57966	3.57966
0.4	2.82478	2.82477

As one might expect, the series approximation begins to lose accuracy for larger T values. Thus, to maintain accuracy a small value of T, T_0, must be chosen to start our integration. However, if T_0 is too small we then begin the integration in a region where r is changing very rapidly. A rule of thumb is to choose T_0 such that $r(T_0) = 0(1)$. For our calculation, values of T_0 = 0.3, 0.4, and 0.5 were used to test the loss of accuracy in the final result by using successively worse initial conditions.

Using the initial value problem described in the preceding section, the results of Table 2 were obtained. The exact solution, as previously given in Eq. (6.8) is given for comparison.

Table 2:

The Function f for Various Values of T_0

t	Exact f(t)	f(t) $T_0 = 0.3$	f(t) $T_0 = 0.4$	f(t) $T_0 = 0.5$
0.1	−0.0920630			
0.2	−0.1613803			
0.3	−0.2096884	−0.2096884		
0.4	−0.2381975	−0.2381975	−0.2382028	
0.5	−0.2476219	−0.2476219	−0.2476261	−0.2476469
0.6	−0.2381975	−0.2381975	−0.2382008	−0.2382169
0.7	−0.2096884	−0.2096884	−0.2096908	−0.2097025
0.8	−0.1613803	−0.1613803	−0.1613819	−0.1613895
0.9	−0.0920630	−0.0920630	−0.0920638	−0.0920676

The solution of the original partial differential
equation (6.1) was computed using the approximation (6.4) and
the values of f(x) given above. For comparison purposes,
the first 150 terms of the Fourier series (6.2) were summed
and, in addition, the Rayleigh-Ritz approximation $w(x, y) =$
$-x(1 - x)(y^2 - 1)$ was computed. The results are given in
Table 3.

Table 3:

Solution of Poisson's Equation by Three Methods

(x,y)	Fourier Series	Kantorovich	Rayleigh-Ritz
(·2,·6)	0.1157170	0.1032834	0.1024000
(·4,·2)	0.2145170	0.2286696	0.2304000
(·6,·6)	0.1688134	0.1524464	0.1536000
(·8,·2)	0.1442439	0.1549251	0.1536000

As a matter of curiosity, the numerical value of the variational integral (6.3) was calculated using both the Rayleigh-Ritz approximation above and the Kantorovich approximation using the function f(x). The results were

$$J_{Rayleigh-Ritz} = -\frac{4}{9} ,$$

$$J_{Kantorovich} = -0.44470199.$$

Thus, the Kantorovich procedure actually does succeed in diminishing the value of J over a similar type of Rayleigh-Ritz approximation.

9. GREEN'S FUNCTIONS

An important tool in the analytical investigation of self-adjoint linear two-point boundary value problems is the

Green's function. Due to the difficulty in obtaining Green's

functions, either analytically or computationally, they are

viewed as of theoretical rather than practical importance.

The aim of this section is to demonstrate how to com-

pute Green's functions as the solution of a stable initial

value problem. As above, the basic idea is to study the

behavior of the Green's function at a fixed point as a func-

tion of the interval length. This idea, of course, may be

viewed as another link in a chain of thought originally forged

by Hadamard.

Consider the differential equation

$$\ddot{x}(t) - z(t)x(t) = -g(t), \qquad 0 < t < T, \tag{1}$$

with the boundary conditions

$$x(0) = 0, \quad \dot{x}(T) = 1. \tag{2}$$

The solution, assumed to uniquely exist, can be represented

in the form

$$x(t) = \int_0^T G(t, \xi)g(\xi) \, d\xi, \tag{3}$$

where the kernel $G(t, \xi)$ is the Green's function of the

problem. Viewed as a function of t, G is a solution of

the problem

$$G - z(t)G = 0, \qquad t \neq \xi, \qquad 0 < t < T, \tag{4}$$

$$G(0, \xi) = 0, \qquad \dot{G}(T, \xi) = 0. \tag{5}$$

In addition, G is continuous at $t = \xi$, but its first derivative has a jump there given by

$$\dot{G}(\xi^+, \xi) - \dot{G}(\xi^-, \xi) = -1. \tag{6}$$

In principle, these properties permit the function G to be constructed as follows. Let $h_0(t)$ be the function such that

$$h_0''(t) - z(t)h_0(t) = 0, \tag{7}$$

$$h_0(0) = 0, \qquad \dot{h}_0(0) = 1. \tag{8}$$

Then on the interval $0 \leq t \leq \xi$, G has the form

$$G(t, \xi) = A h_0(t), \tag{9}$$

where A is independent of t. Let $h_1(t)$ be the solution of the problem

$$h_1''(t) - z(t)h_1(t) = 0, \tag{10}$$

$$h_1(T) = 1, \qquad \dot{h}_1(T) = 0. \tag{11}$$

On the interval $\xi \le t \le T$, G has the form

$$G(t, \xi) = Bh_1(t), \tag{12}$$

where B is also independent of t. The parameters A and B are determined by the continuity and jump conditions at $t = \xi$,

$$Ah_0(\xi) - Bh_1(\xi) = 0, \tag{13}$$

$$A\dot{h}_0(\xi) - B\dot{h}_1(\xi) = 1. \tag{14}$$

Thus, the Green's function is

$$G(t, \xi) = \begin{cases} -h_1(\xi)h_0(t)/\Delta, & 0 \le t \le \xi \\ -h_0(\xi)h_1(t)/\Delta, & \xi \le t \le T, \end{cases} \tag{15}$$

where

$$\Delta = h_0(\xi)\dot{h}_1(\xi) - \dot{h}_0(\xi)h_1(\xi)$$

$$= \text{constant}. \tag{16}$$

Incidentally, this shows that G is symmetric, i.e., $G(t, \xi) = G(\xi, t)$.

The procedure just outlined shows the structure of the Green's function but, in many cases, would be difficult to carry out numerically since the production of the functions h_0 and h_1 requires the integration of the equation

$$h'' - z(t)h(t) = 0 \tag{17}$$

in the directions of both increasing and decreasing t. If

this equation has both increasing and decreasing exponential

solutions (e.g., when $z(t) = k^2$, k constant), it may be

difficult to accurately carry out the integrations due to

growth of roundoff errors, a well known phenomenon. In the

next section we shall employ our initial value approach to

combat this difficulty.

10. A CAUCHY SYSTEM FOR G

Utilizing the Cauchy system in Section 3, we may now

derive an initial value problem for the function G. We begin

by rewriting Eq. (7.1) in the form

$$\dot{x}(t, T) = y, \quad x(0, T) = 0, \tag{1}$$

$$\dot{y}(t, T) = z(t)x - g(t), \quad y(T, T) = 1,$$
$$0 \leq t \leq T. \tag{2}$$

In terms of the general linear system treated previously, we

identify the functions as $a = 0$, $b = 1$, $c = z(t)$, $d = 0$, $f =$

$-g(t)$, $\alpha_1 = \alpha_4 = 1$, $\alpha_2 = \alpha_3 = 0$. The appropriate Cauchy

problem is

$$r'(T) = 1 - z(T)r^2, \quad r(0) = 0, \tag{3}$$

$$m'(T) = -[mz(T) - g(T)]r, \quad m(0) = 0, \tag{4}$$

$$u_T(t, T) = -[mz(T) - g(T)]p(t, T), \quad u(t, t) = m(t), \tag{5}$$

$$p_T(t, T) = -r(T)z(T)p(t, T), \quad p(t, t) = r(t). \tag{6}$$

Since we have no need for them, we disregard the functions v and q. We also note that the form of the boundary conditions implies that the auxiliary functions $s = 1$, $n = 0$.

Standard results in the theory of differential equations tell us that

$$u(t, T) = \int_0^T G(t, \xi, T)g(\xi) \, d\xi, \tag{7}$$

where we now explicitly indicate the dependence of G upon T as well as t and ξ. Differentiating this formula with respect to T gives

$$u_T(t, T) = G(t, T, T)g(T) + \int_0^T G_T(t, \xi, T)g(\xi) \, d\xi. \tag{8}$$

Furthermore,

$$u(T, \ T) = m(T) = \int_0^T G(T, \ \xi, \ T)g(\xi) \ d\xi. \qquad (9)$$

Comparing Eqs. (5) and (8) while making use of (9), it follows

by uniqueness that

$$p(t, \ T) = G(t, \ T, \ T) \qquad (10)$$

and that

$$G_T(t, \ \xi, \ T) = -z(T)p(t, \ T)G(T, \ \xi, \ T). \qquad (11)$$

Consequently, G satisfies the equation

$$G_T(t, \ \xi, \ T) = -z(T)G(t, \ T, \ T)G(T, \ \xi, \ T) \qquad (12)$$

or, upon making use of the symmetry of G,

$$G_T(t, \ \xi, \ T) = -z(T)p(t, \ T)p(\xi, \ T). \qquad (13)$$

Eq. (12) is an Hadamard variational formula, whereas the

numerical procedure uses Eq. (13) in which p is regarded as

a forcing term.

11. A NUMERICAL EXAMPLE

The numerical procedure is based upon the equations

for G, p, and r. First, the equation for r is numerically integrated from T = 0 to T = min(t, ξ). For definiteness,

assume t < ξ. Thus, the numerical value of r(t) is known.

At T = t the equation for p(t, T) is adjoined, and the

integration is continued from T = t to T = ξ. At this

point the two equations for p(ξ, T) and G(t, ξ, T) are

adjoined, and the integration is continued to T = T*, the

desired interval length. The numerical value of G(t, ξ, T*)

is the desired Green's function.

   To demonstrate the utility of this approach, consider

the problem

$$\ddot{x} - k^2 x = -g(t), \qquad 0 < t < T \tag{1}$$

$$x(0) = 0, \quad \dot{x}(T) = 0, \tag{2}$$

where k is a constant. The Green's function is

$$G(t, \xi, T) = \begin{cases} \dfrac{1}{k} \dfrac{\cosh k(\xi-T) \sinh kt}{\cosh k\,T}, & 0 \le t \le \xi, \\[3mm] \dfrac{1}{k} \dfrac{\sinh k\xi \cosh k(t-T)}{\cosh k\,T}, & \xi \le t \le T. \end{cases}$$

Using an Adams-Moulton integration procedure with step size

0.01, the above procedure was employed to calculate

G(t, ξ, T)(T ≤ 1) for k = 10 at a grid spacing of 0.1 in

the (t,ξ)-plane. Five significant figure accuracy was

obtained in a matter of a few seconds of computing time.

12.   NONLINEAR TWO-POINT BOUNDARY VALUE PROBLEMS

Frequently, a major step in the solution of many problems is the numerical treatment of a nonlinear two-point boundary value problem. Problems of this sort naturally arise in optimal control theory and in other types of variational problems as well as in various parts of astrophysics and biology. We shall return to these topics in a later chapter but, for the moment, let us consider only the mathematical aspects of a class of nonlinear two-point boundary value problems to demonstrate the application of the ideas developed above for the linear case.

As one might expect in dealing with nonlinear problems, the lack of the principle of superposition forces us to lower our sights somewhat in the development of our initial value representation. We can no longer expect to remain within the realm of ordinary differential equations since, in general, the solution will now be a function of the prescribed boundary conditions as well as the interval length. This immediately pushes us into the realm of partial differential equations to account for these additional variables. However, as will be seen, the types of partial differential equations occurring are of a very simple type which lend themselves to analysis and numerical solution by a number of

techniques. As a result, we can confidently expect to use
the Cauchy representation to solve a wide variety of nonlinear problems.

We consider the system of nonlinear ordinary differential equations

$$\dot{u} = F(u, v, t), \tag{1}$$

$$\dot{v} = G(u, v, t), \qquad 0 < t < T, \tag{2}$$

subject to the boundary conditions

$$u(0) = 0, \tag{3}$$

$$v(T) = c. \tag{4}$$

For the sake of exposition, we assume u and v are scalar
functions. The multidimensional versions of our results can
be readily obtained with a modicum of effort. To indicate
the dependence of the functions u and v upon c and T,
as well as upon t, we shall occasionally write $u(t, c, T)$
and $v(t, c, T)$ when necessary for clarity.

By differentiating Eqs. (1)-(4) with respect to c,
it is seen that

$$\dot{u}_c(t, c, T) = F_u u_c + F_v v_c, \tag{5}$$

$$\dot{v}_c(t, c, T) = G_u u_c + G_v v_c, \qquad 0 < t < T, \tag{6}$$

$$u_c(0, c, T) = 0, \tag{7}$$

$$v_c(T, c, T) = 1. \tag{8}$$

Similarly, a differentiation in T yields

$$\dot{u}_T = F_u u_T + F_v v_T, \tag{9}$$

$$\dot{v}_T = G_u u_T + G_v v_T, \qquad 0 < t < T, \tag{10}$$

$$u_T(0, c, T) = 0, \tag{11}$$

$$\dot{v}(T, c, T) + v_3(T, c, T) = 0. \tag{12}$$

In the above equation, $\dot{v}$ is the derivative of v with respect to the first argument, while v_3 is the derivative with respect to the third argument.

To make use of these equations, note that from the differential equation (2), when $t = T$ we have

$$\dot{v}(T, c, T) = G(u(T, c, T), v(T, c, T), T)$$

$$= G(r(c, T), c, T) \tag{13}$$

where the notation

$$r(c, T) = u(T, c, T) \tag{14}$$

has been introduced. Comparing Eqs. (5)-(8) with (9)-(12), and assuming a unique solution exists, it follows that

$$u_T(t, c, T) = -G(r(c, T), c, T)u_c(t, c, T), \tag{15}$$

$$v_T(t, c, T) = -G(r(c, T), c, T)v_c(t, c, T),$$

$$0 \le t \le T, \quad |c| < \infty. \tag{16}$$

Equations (15) and (16) are the desired partial differential equations for u and v. The initial conditions at $T = t$ are

$$u(t, c, t) = r(c, t), \tag{17}$$

$$v(t, c, t) = c. \tag{18}$$

It remains to consider the function r.

Differentiate Eq. (14) with respect to T to obtain

$$r_T(c, T) = \dot{u}(T, c, T) + u_3(T, c, T). \tag{19}$$

From Eqs. (1) and (15), we now see that

$$r_T(c, T) = F(r(c, T), c, T) -$$

$$G(r(c, T), c, T)r_c(c, T). \tag{20}$$

This is the quasilinear first-order partial differential equation satisfied by r. From Eq. (3) we see that

$$r(c, 0) = 0. \tag{21}$$

The equations for u, v, and r, together with their initial conditions, constitute the initial value representation for the original nonlinear problem.

13. STATEMENT OF THE INITIAL VALUE PROBLEM

For sufficiently small $T > 0$, we assume the function r is determined as the solution of the Cauchy problem

$$r_T = F(r, c, T) - G(r, c, T)r_c, \qquad T > 0, \tag{1}$$

$$r(c, 0) = 0. \tag{2}$$

The function u is the solution of the problem

$$u_T = G(r, c, T)u_c, \qquad t < T, \tag{3}$$

$$u(t, c, t) = r(c, T), \tag{4}$$

while the function v satisfies

$$v_T = G(r, c, T)v_T, \qquad t < T, \tag{5}$$

$$v(t, c, t) = c. \tag{6}$$

Again we note that only integration in the direction of increasing T is required. Thus, the procedure is a one-sweep method.

14. COMPUTATIONAL METHODS

Let us briefly indicate a few of the methods which may be used to numerically treat the above first-order partial differential equations. For more complete details on these procedures, the reader is referred to the references cited at the conclusion of the chapter.

The standard numerical technique used for dealing with partial differential equations is the method of finite differences. Although there are almost as many finite difference schemes as there are investigators, the underlying idea in all procedures is to replace the partial derivatives by finite differences and then deal with the resulting algebraic problem.

To illustrate, let us consider Eq. (13.1) for the function r,

$$r_T(c, \ T) = F(r, \ c, \ T) - G(r, \ c, \ T)r_c, \qquad (1)$$

$$r(c, \ 0) = 0. \qquad (2)$$

The simplest type of difference scheme is the single step

forward difference procedure, by which we write

$$r_T(c, T) \cong \frac{r(c,T+\Delta) - r(c,T)}{\Delta} , \tag{3}$$

$$r_c(c, T) = \frac{f(c+\delta,T) - r(c,T)}{\delta} , \quad \Delta, \quad \delta \ll 1. \tag{4}$$

Substituting these approximations into (1) and re-arranging

terms gives

$$r(c, T + \Delta) = r(c, T) + \Delta\Big\{F(r, c, T) -$$

$$G(r, c, T) \Big[\frac{r(c+\delta,T) - r(c,T)}{\delta}\Big]\Big\}, \tag{5}$$

$$r(c, 0) = 0. \tag{6}$$

Equation (5), together with the line of data specified by

(6), allows us to compute r on the line $T + \Delta$ in terms

of its values on the line T. Under suitable hypotheses on

F, G and ratio Δ/δ, it can be established that as $\Delta, \delta \to 0$,

the difference approximation converges to the solution of the

original problem.

The approximations given by (3) and (4) are, in

general, only accurate to terms of order Δ or δ for most

purposes. This is not a particularly good approximation and

recourse must be taken to more accurate schemes. There are a number of schemes which, for very modest increase in comput-

ing burden, give much improved accuracy. For example, a

central difference procedure which approximates r_T and r_c

as

$$r_T(c, T) \stackrel{\sim}{=} \frac{r(c,T+\Delta) - r(c,T-\Delta)}{2\Delta} , \qquad (7)$$

$$r_c(c, T) \stackrel{\sim}{=} \frac{r(c+\delta,T) - r(c-\delta,T)}{2\delta} , \qquad (8)$$

is accurate to order Δ^2 and δ^2, respectively. In a similar

manner, higher order schemes can be constructed. As in the

central difference case, the higher order schemes usually

need more than one line of data to get started. Consequently,

it is usually necessary to employ a lower order procedure to

initialize the more accurate schemes. Examples of this are

given in the references at the end of the chapter.

Another approach to the solution of (1) is through

power series expansions. Although series expansions have

been used for many years as an analytical tool in differential

equations, their computational use has generally been limited

due to factors such as limited regions of convergence and

slow rates of convergence. However, for problems such as

(1), particularly when the functions F and G are poly-

nomial functions of their arguments, there are a number of

ways to accelerate the convergence and expand the domain of

applicability of a power series representation for the func-

tion   r.   This solution approach is particularly applicable

when the function   r   is desired for only a small number of

c   and   T   values.

To illustrate the use of power series, let the func-

tion   r   be expressed as

$$r(c, T) = \sum_{i=0}^{\infty} \sum_{j=0}^{\infty} a_{ij} c^i T^j. \tag{9}$$

Furthermore, assume that the functions   F   and $r_c G$ can be

written as the power series

$$F(r, c, T) = \sum_{i=0}^{\infty} \sum_{j=0}^{\infty} b_{ij} c^i T^j, \tag{10}$$

$$r_c G(r, c, T) = \sum_{i=0}^{\infty} \sum_{j=0}^{\infty} d_{ij} c^i T^j, \tag{11}$$

where the numbers $\{b_{ij}\}$ and $\{d_{ij}\}$ are dependent upon the

$\{a_{ij}\}$ through the specific forms of   F   and   G.   Performing

the required partial differentiations upon (9), and substitu-

ting the results into Eq. (1) gives

$$\sum_{i=0}^{\infty} \sum_{j=0}^{\infty} (j + 1) a_{i,j+1} c^i T^j = \sum_{i=0}^{\infty} \sum_{j=0}^{\infty} b_{ij} c^i T^j$$

$$-\sum_{i=0}^{\infty} \sum_{j=0}^{\infty} d_{ij} c^i T^j. \tag{12}$$

Equating coefficients of $c^i T^j$, we have the recurrence relation

$$(j + 1)a_{i,j+1} = b_{ij} - d_{ij}, \qquad i,j = 0,1,2,\dots . \tag{13}$$

The initial condition obtained from Eq. (2) is

$$a_{i0} = 0, \qquad i = 0,1,2,\dots . \tag{14}$$

Using Eqs. (13) and (14), a digital computer can easily generate the expansion coefficients $\{a_{ij}\}$ in a matter of moments. Truncating the series (9) at suitably large values of i and j, an accurate solution for r may be obtained for $|c| + |T|$ sufficiently small. If (9) possesses singularities for certain values of c or T, analytic continuation procedures must be employed.

A third method for the solution of (1) is a type of finite difference scheme termed differential quadrature. In this procedure, the derivatives with respect to all but one of the independent variables are replaced by suitable linear combinations of function values. This reduces the problem to a set of ordinary differential equations with known initial values.

For example, let us approximate the term r_c at a set of points $\{c_i\}$ by

$$r_c(c_i, T) \cong \sum_{j=1}^{N} a_{ij} r(c_j, T). \tag{15}$$

For simplicity, assume $0 \le c_i \le 1$, $i = 1,2,\ldots,N$ and let us determine $\{a_{ij}\}$ such that (15) is exact for polynomials of degree less than N. Utilizing the test functions c^i, $i = 1,2,\ldots,N$, inversion of a Vandermonde matrix yields the correct expansion numbers. To avoid matrix inversions, the

test polynomials $r(c, T) = \dfrac{P_N^*(c)}{(c-c_i)P_N^{*'}(c_i)}$, $i = 1,2,\ldots,N$,

may be employed. Here P_N^* is the $N\underline{th}$ shifted Legendre polynomial, c_i is its $i\underline{th}$ root. The set $\{P_N^*\}$ is orthogonal on $[0, 1]$ and the expansion numbers $\{a_{ij}\}$ may be directly obtained in terms of the roots of P_N^* and $P_N^{*'}$ evaluated at these roots. The resulting expressions are

$$a_{ik} = \begin{cases} \dfrac{P_N^{*'}(c_i)}{(c_i-c_k)P_N^{*'}(c_k)} & , \quad i \ne k, \\[4ex] \dfrac{1-2c_k}{2c_k(c_k-1)} & , \quad i = k. \end{cases} \tag{16}$$

Adopting the notation,

$$r(c_i, T) = r_i(T), \quad i = 1,2,\ldots,N, \tag{17}$$

and employing the approximation (15) in (1), we obtain the set of ordinary differential equations

$$r_i'(T) = F(r_i, c_i, T) - G(r_i, c_i, T) \sum_{j=1}^{N} a_{ij} r_j(T),$$

$$i = 1,2,\ldots,N. \tag{18}$$

The initial conditions are

$$r_i(0) = 0. \tag{19}$$

Equations (18)-(19) comprise a nonlinear set of ordinary differential equations suitable for computing the functions $r_i(T)$.

15. VALIDATION OF THE INITIAL VALUE PROBLEM

At this point it is important to establish the validity of the Cauchy system presented in section 13 by proving its equivalence with the original nonlinear two-point boundary value problem. Our basic assumption will be that the interval length T is sufficiently short and the functions F and G sufficiently smooth that the Cauchy system possesses a unique solution.

First consider Eq. (13.3) with the initial condition (13.4). Differentiating both sides of (13.3) with respect to t gives

$$\dot{u}_T(t, \text{ c}, \text{ T}) = G(r, \text{ c}, \text{ T})\dot{u}_c(t, \text{ c}, \text{ T}), \qquad t < T. \qquad (1)$$

Differentiation of Eq. (13.4) with respect to t yields

$$r_t(c, \text{ t}) = \dot{u}(t, \text{ c}, \text{ t}) + u_3(t, \text{ c}, \text{ t}) =$$

$$\dot{u}(t, \text{ c}, \text{ t}) + G(r(c, \text{ t}), \text{ c}, \text{ t})r_c(c, \text{ t}). \qquad (2)$$

By Eq. (13.1),

$$r_t(c, \text{ t}) = F(r(c, \text{ t}), \text{ c}, \text{ t}) - G(r(c, \text{ t})c, \text{ t})r_c(c, \text{ t}). \qquad (3)$$

It follows that

$$\dot{u}(t, \text{ c}, \text{ t}) = F(r(c, \text{ t}), \text{ c}, \text{ t}). \qquad (4)$$

Equations (1) and (4) form an initial value problem for the function $\dot{u}(t, \text{ c}, \text{ T})$.

Next consider the function $F(u, \text{ v}, \text{ t})$. By differentiation it is seen that

$$F_T = F_u u_T + F_v v_T, \qquad (5)$$

$$F_c = F_u u_c + F_v v_c, \tag{6}$$

$$F_T - GF_c = F_u[u_T - Gu_c] + F_v[v_T - Gv_c], \tag{7}$$

which, through Eqs. (13.3) and (13.5), implies

$$F_T - GF_c = 0. \tag{8}$$

Moreover, for $T = t$,

$$F(u(t, c, t), v(t, c, t), t) = F(r(c, t), c, t). \tag{9}$$

Thus, we see that the functions $\dot{u}(t, c, T)$ and $F(u(t, c, T), v(t, c, T), t)$ satisfy the same initial value problem for $T \geq t$. By our uniqueness assumption, it follows that

$$\dot{u}(t, c, T) = F(u(t, c, T), v(t, c, T), t). \tag{10}$$

In a similar manner, the functions $\dot{v}(t, c, T)$ and $G(u(t, c, T), v(t, c, T, t)$ are seen to satisfy the same differential equation.

It remains to consider the boundary conditions on the functions u and v. The function v is constructed so that $v(T, c, T) = c, T > 0$. Lastly, it will be shown that

$$u(0, c, T) = 0. \tag{11}$$

The function $u(0, c, T)$ satisfies the partial differential equation

$$u_T(0, c, T) = G(r(c, T), c, T)u_c(0, c, T),$$

$$T > 0, \tag{12}$$

along with the initial condition

$$u(0, c, 0) = r(c, 0) = 0. \tag{13}$$

Thus, by uniqueness, $u(0, c, T) \equiv 0$ is the only solution of (12)–(13) which completes our validation.

CHAPTER THREE

NOTES AND REFERENCES

§2.   A geometric derivation of the initial value problem

        for a more restricted type of boundary condition is

        given in

        Bellman, R., H. Kagiwada, and R. Kalaba, "Invariant

        Imbedding and the Numerical Integration of Boundary

        Value Problems for Unstable Linear Systems of

        Ordinary Differential Equations", Comm. ACM, 10

        (1967), 100-102.

        Related results for initial value problems together

        with numerical examples, are given in

        Scott, M., "Numerical Solutions of Unstable Initial

        Value Problems by Invariant Imbedding", Sandia

        Laboratories, Research Rpt. SC-RR-69-343, August

        1969.

        Numerous examples from the areas of chemical engineer-

        ing and estimation theory may be found in

        Lee, E. S., Quasilinearization and Invariant

        Imbedding, Academic Press, New York 1968.

§4.   Classic references on stability theory of ordinary

        differential equations are

Bellman, R., The Stability Theory of Differential
Equations, McGraw-Hill, New York, 1953.

LaSalle, J. and S. Lefschetz, Stability by
Liaponov's Direct Method with Applications,
Academic Press, New York, 1961.

For a discussion of numerical rather than analytical
instability, see

Babuska, I., M. Prager, and E. Vitasek, Numerical
Processes in Differential Equations, Interscience,
New York, 1966.

Collatz, L., Numerical Treatment of Differential
Equations, Springer-Verlag, Berlin, 1960.

§6.   For more detailed results and examples of
Kantorovich's extension of the Rayleigh-Ritz procedure
see

Kantorovich, L. and V. Krylov, Approximate Methods
of Higher Analysis, Interscience, New York, 1958.

§9.   Numerical examples using a related technique are
given in

Huss, R. and R. Kalaba, "Invariant Imbedding and
the Numerical Determination of Green's Functions",
to appear J. Opt. Th. Appl.

Related results on Green's functions and characteristic values are found in

Kagiwada, H., R. Kalaba, A. Schumitzky, and R. Sridhar, "An Integral Equation and a Representation for a Green's Function", J. Math. Anal. Appl. 2 (1968), 226-229.

Scott, M., L. Shampine, and G. Wing, "Invariant Imgedding and the Calculation of Eigenvalues for Sturm-Liouville Systems", Computing, 4 (1969), 10-23.

§12.   These results were first given in

Kagiwada, H. and R. Kalaba, "Derivation and Validation of an Initial Value Method for Certain Nonlinear Two-Point Boundary Value Problems", J. Opt. Th. Appl. 6 (1968), 378-385.

§14.   Detailed discussions of various finite difference schemes are to be found in

R. Richtmyer and K. Morton, Difference Methods for Initial Value Problems, Interscience, New York, 1957.

See also the book by Collatz cited under §4 above.

Power series treatments of initial value and boundary

value problems are discussed in

Casti, J., "Power Series and the Numerical Treatment of Initial Value Problems", The RAND Corp.,
RM-6059-PR, September 1969.

Leavitt, J., "A Power Series Method for Solving Nonlinear Boundary Value Problems", Quart. Appl. Math.
27 (1969), 67-77.

Differential quadrature methods with numerical exam

examples are presented in

Bellman, R., J. Casti, and B. Kashef, "Differential
Quadrature: A New Technique for the Rapid Solution
of Nonlinear Partial Differential Equations",
J. Comp. Physics, 10 (1972), 40-52.

CHAPTER FOUR

FREDHOLM INTEGRAL EQUATIONS

1. INTRODUCTION

In view of the success of the invariant imbedding method in the attack upon two-point boundary value problems and the well-known connections between boundary-value problems and Fredholm integral equations, our next objective will be to test our procedures upon various types of linear integral equations. The focal point of these investigations will be the integral equation of convolution type

$$u(t) = g(t) + \int_0^c k(|t - y|)u(y) \, dy, \qquad (1)$$

which occurs in optimal filtering theory, polymer chemistry, radiative transfer, and many other branches of physics and engineering.

We begin by demonstrating that by introduction of a suitable imbedding parameter, we may obtain the solution of Eq. (1) in terms of an auxiliary function Φ. During the course of obtaining the representation formula for the function u in terms of Φ, some non-classical results concerning the Fredholm resolvent are obtained.

Next a numerical method suitable for calculating the solutions of Eq. (1) using the imbedding equations is presented, along with calculations for a specific example.

Our next order of business is the analytic proof that the Cauchy problem presented actually does solve Eq. (1). As the proof proceeds, several new functions are introduced which are of interest in their own right.

Finally, the chapter concludes with a discussion of several topics in integral equations and their relations to the imbedding theory. We see that characteristic functions and values may be obtained from the imbedding equations and that it is also possible to calculate the Fredholm resolvent by means of a Cauchy problem. The intimate connections between the imbedding formalism and these well-studied areas of classical analysis open the way for a much deeper understanding of the invariant imbedding formalism.

2. THE BASIC PROBLEM

In order to present our ideas in as simple a fashion as possible, but cover a broad class of nontrivial and important problems, we confine our attention to the Fredholm integral equation with convolution kernel

$$u(t) = g(t) + \int_0^c k(|t - y|)u(y) \, dy, \tag{1}$$

where it is assumed that the kernel k is expressible in the form

$$k(r) = \int_0^1 e^{-r/z'} w(z') \, dz', \quad r > 0. \tag{2}$$

Furthermore, we assume that the function $w(z)$ is such that Eq. (1) has a unique solution on the interval $0 \leq t \leq c$ for all continuous forcing terms $g(t)$.

3. AN IMBEDDING

By analogy with the treatment of boundary value problems, we seek to imbed Eq. (2) within a family of problems for which it is possible to derive relations linking adjacent members of the family and for which a suitable "initial condition" may be found.

4.   A REPRESENTATION FORMULA

To obtain the solution of the family (3.1), we
invoke a powerful, conceptual tool of analysis, the representation formula. We will attempt to obtain a function
which can be calculated, and in terms of which the solution
to the family of integral equations may be expressed.

Let the function $\Phi(t, x)$ be defined by the integral
equation

$$\Phi(t, x) = k(x - t) + \int_0^x k(|t - y|)\Phi(y, x) \, dy,$$

$$0 \leq t \leq x. \tag{1}$$

Also, define the function f by the equation

$$f(x) = g(x) + \int_0^x \Phi(y, x)g(y) \, dy, \quad x \geq 0. \tag{2}$$

Then we claim that the solution of the family of integral
equations

$$u(t, x) = g(t) + \int_0^x k(|t - y|)u(y, x) \, dy,$$

$$x \geq t \tag{3}$$

is given by the formula

$$u(t, x) = f(t) + \int_t^x \Phi(t, y)f(y)\, dy,$$

$$0 \le t \le x. \tag{4}$$

The proof proceeds very simply. Differentiate both sides of Eq. (3) with respect to x to obtain

$$u_x(t, x) = k(x - t)u(x, x) +$$

$$\int_0^x k(|t - y|)u_x(y, x)\, dy. \tag{5}$$

Regarding Eq. (5) as an integral equation for the function $u_x(t, x)$, we see that the solution is

$$u_x(t, x) = \Phi(t, x)u(x, x), \qquad 0 \le t \le x. \tag{6}$$

Through integration this last equation becomes

$$u(t, x) = u(t, t) + \int_t^x \Phi(t, y)u(y, y)\, dy. \tag{7}$$

For $u(x, x)$ we may write

$$u(x, x) = g(x) + \int_0^x k(x - y)u(y, x)\, dy. \tag{8}$$

Cross-multiplication of Eqs. (1) and (3), integration on t

from $t = 0$ to $t = x$ and use of the symmetry of the kernel k, gives

$$\int_0^x \Phi(t,\ x)g(t)dt = \int_0^x u(t,\ x)k(x - t)\ dt. \tag{9}$$

Thus, Eq. (8) becomes

$$u(x,\ x) = g(x) + \int_0^x \Phi(t,\ x)g(t)\ dt, \quad x \geq 0. \tag{10}$$

Equation (7) then assumes the form

$$u(t,\ x) = g(t) + \int_0^t \Phi(t',\ t)g(t')\ dt' +$$

$$\int_t^x \Phi(t,\ y)[g(y) + \int_0^y \Phi(t',\ y)g(t')\ dt']\ dy$$

$$= f(t) + \int_t^x \Phi(t,\ y)f(y)\ dy, \tag{11}$$

which was to be established.

5. THE BELLMAN-KREIN EQUATION

The above representation formula gives a means for obtaining a very elegant derivation of the Bellman-Krein formula for the resolvent kernel of Eq. (4.3).

We recall that the resolvent K yields the solution

of Eq. (4.3) in the form

$$u(t, x) = g(t) + \int_0^x K(t, y, x)g(y) \, dy,$$

$$0 \leq t \leq x. \tag{1}$$

It is known that the resolvent satisfies the integral
equation

$$K(t, y, x) = k(|t - y|) +$$

$$\int_0^x k(|t - y'|)K(y', y, x) \, dy'. \tag{2}$$

Comparing Eq. (2) with the earlier Eq. (4.1) for Φ, we see
that

$$\Phi(t, x) = K(t, x, x) = K(x, t, x), \tag{3}$$

where we have used the known symmetry relationship

$$K(t, y, x) = K(y, t, x). \tag{4}$$

To apply our representation formula, let

$$g(t) = k(|t - y'|) \tag{5}$$

in Eq. (4.3). It is known that the solution u is the
resolvent

$$u(t, x) = K(t, y', x).   \tag{6}$$

Use of the representation formula, Eq. (4.11), shows that

$$K(t, y, x) = \Phi(y, t) + \int_t^x \Phi(t, z)\Phi(y, z) \, dz.$$

$$0 \le y \le t \le x.   \tag{7}$$

Differentiation with respect to x then provides the Bellman-Krein formula

$$K_x(t, y, x) = \Phi(t, x)\Phi(y, x),   \tag{8}$$

which expresses the change in the resolvent with changes in the interval length.

The Bellman-Krein equation plays an important role in many theoretical investigations in integral equations and, as we shall see, proves extremely useful as an aid to the development of our Cauchy problem.

6.   DETERMINATION OF $\Phi(t, x)$

The representation formula for $u(t, x)$ shows that knowledge of the function $\Phi(t, x)$ is all that is necessary in order to obtain the solution to the family of integral equations. In order to obtain useful equations for Φ, we introduce the new functions $J(t, x, z)$, $X(x, z)$, and $Y(x, z)$

by the relations

$$J(t, x, z) = e^{-(x-t)/z} +$$

$$\int_0^x k(|t - y|)J(y, x, z) \, dy,$$

$$x \geq t, \qquad 0 \leq z \leq 1, \quad (1)$$

$$X(x, z) = J(x, x, z), \quad x \geq 0, \qquad 0 \leq z \leq 1, \qquad (2)$$

$$Y(x, z) = J(0, x, z), \quad x \geq 0, \qquad 0 \leq z \leq 1. \qquad (3)$$

In view of the analytic form of the kernel as given

in Eq. (2.2), multiplication of Eq. (1) by the function $w(z)$

and then integration of both sides from $z = 0$ to $z = 1$

yields the fundamental relation

$$\Phi(t, x) = \int_0^1 J(t, x, z')w(z') \, dz', \quad x \geq t. \qquad (4)$$

Thus, we see that, knowledge of the function $J(t, x, z)$

suffices to determine Φ.

Our immediate goal will be to obtain a Cauchy problem

for the function J. As will become evident in the sequel,

the functions X and Y will play a role.

7.   AN INITIAL VALUE PROBLEM FOR   J

We begin the derivations by differentiation of Eq. (6.1) with respect to x, obtaining

$$J_x(t, x, z) = -\frac{1}{z} e^{-(x-t)/z} + k(x - y)J(x, x, z) +$$

$$\int_0^x k(|t - y|)J_x(y, x, z) \, dy, \quad x \geq t,$$

$$0 \leq z \leq 1. \quad (1)$$

Regarding Eq. (1) as an integral equation for the function $J_x(t, x, z)$ and recalling Eqs. (6.1) and (4.1) for the functions J and Φ, we may write J_x as

$$J_x(t, x, z) = -\frac{1}{z} J(t, x, z) + X(x, z)\Phi(t, x). \quad (2)$$

Using Eq. (6.4), the last equation becomes

$$J_x(t, x, z) = -\frac{1}{z} J(t, x, z) +$$

$$X(x, z)\int_0^1 J(t, x, z')w(z') \, dz'. \quad (3)$$

Through the definition of the function X(x, z) in Eq. (6.2), it is seen that an appropriate initial condition for Eq. (3) is

$$J(t, t, z) = X(t, z). \tag{4}$$

The function J will be determined as soon as a means for determining X is available. This task will occupy our attention for the next few sections.

8. THE FIRST DERIVATION OF THE EQUATION FOR X

By definition, the function $X(x, z)$ is

$$X(x, z) = J(x, x, t) = 1 + \int_0^x k(x - y)J(y, x, z) \, dy,$$

$$x \geq 0, \qquad 0 \leq z \leq 1. \tag{1}$$

Let us adopt the notation $J_1(t, x, z)$ for the derivative of J with respect to its first argument. Then differentiating Eq. (1) with respect to x gives

$$\frac{d}{dx} X(x, z) = \frac{d}{dx} J(x, x, z) =$$

$$[J_1(t, x, z) + J_x(t, x, z)]_{t=x}. \tag{2}$$

However, we have already obtained the function $J_x(t, x, z)$ in Eq. (7.2), so all that remains is to find $J_1(t, x, z)$.

To do this, write the integral equation for J in the form

$$J(t, x, z) = e^{-(x-t)/z} + \int_0^t k(t - y)J(y, x, z) \, dy +$$

$$\int_t^x k(y - t)J(y, x, z) \, dy. \quad (3)$$

Next, differentiate Eq. (3) with respect to t. This yields

$$J_1(t, x, z) = \frac{1}{z} e^{-(x-t)/z} + k(0)J(t, x, z) +$$

$$\int_0^t \frac{\partial}{\partial t} k(t - y)J(y, x, z) \, dy - k(0)J(t, x, z) -$$

$$\int_t^x \frac{\partial}{\partial t} k(t - y)J(y, x, z) \, dy = \frac{1}{z} e^{-(x-t)/z} +$$

$$\int_0^t \frac{\partial}{\partial t} k(t - y)J(y, x, z) \, dy -$$

$$\int_t^x \frac{\partial}{\partial t} k(y - t)J(y, x, z) \, dy,$$

$$x \geq t, \qquad 0 \leq z \leq 1. \quad (4)$$

Integration of the two integral terms of Eq. (4) by parts gives the relations

$$\int_0^t \frac{\partial}{\partial t} k(t - y)J(y, x, z) \, dy = -k(0)J(t, x, z) +$$

$$k(t)J(0, x, z) + \int_0^t k(t - y)J_1(y, x, z) \, dy, \quad (5)$$

$$\int_t^x \frac{\partial}{\partial t} k(y - t)J(y, x, z) \, dy = -k(0)J(t, x, z) +$$

$$k(x - t)J(x, x, z) - \int_t^x k(y - t)J_1(y, x, z) \, dy. \quad (6)$$

Combining Eqs. (4)-(6) gives the result

$$J_1(t, x, z) = \frac{1}{z} e^{-(x-t)/z} + k(t)J(0, x, z) -$$

$$k(x - t)J(x, x, z) + \int_0^x k(|t - y|)J_1(y, x, z) \, dy. \quad (7)$$

Viewing Eq. (7) as an integral equation for the function J_1, and keeping in mind the equations for Φ, X, and Y, gives the final result

$$J_1(t, x, z) = \frac{1}{z} J(t, x, z) + Y(x, z)\Phi(x - t, x) +$$

$$X(x, z)\Phi(t, x). \quad (8)$$

Returning now to Eq. (2), we may write the final equation for $X(x, z)$ in the form

$$\frac{d}{dx} X(x, z) = \frac{1}{z} J(x, x, z) + Y(x, z)\Phi(0, x) -$$

$$X(x, z)\Phi(x, x) + [-\frac{1}{z} J(x, x, z) + X(x, z)\Phi(x, x)],$$

$$= Y(x, z)\Phi(0, x),$$

$$= Y(x, z) \int_0^1 Y(x, z')w(z') \, dz', \quad x > 0, \tag{9}$$

where use has been made of the definition of the function

$Y(x, z)$ as given in Eq. (6.3).

The Cauchy problem for the function Y may be immediately obtained from Eq. (7.2) by putting $t = 0$. This substitution gives

$$\frac{d}{dx} Y(x, z) = -\frac{1}{z} Y(x, z) + X(x, z)\Phi(0, x),$$

$$= -\frac{1}{z} Y(x, z) + X(x, z) \int_0^1 Y(x, z')w(z') \, dz',$$

$$x > 0. \tag{10}$$

Equations (9) and (10) form a coupled pair of differential-integral equations for the functions X and Y. In view of their definitions, the functions X and Y have as initial conditions at $x = 0$,

$$X(0, z) = Y(0, z) = 1. \tag{11}$$

These equations close out the Cauchy system for the functions

J, X, and Y.

9.   SECOND DERIVATION OF THE EQUATION FOR X

We may take advantage of the special nature of the
convolution kernel to obtain an entirely different derivation
of the Cauchy problem for X.

Since the function X is defined to be J(x, x, z),
we will work with the integral equation for J. Recall that
J satisfies the integral equation

$$J(t, x, z) = e^{-(x-t)/z} + \int_0^x k(|t - y|)J(y, x, z)\, dy,$$

$$x \geq t, \qquad 0 \leq z \leq 1. \quad (1)$$

To begin, make the substitutions $t \to x - t,\ y \to x - y$ in
Eq. (1). This gives the new integral equation for the
function $J(x - t, x, z)$ as

$$J(x - t, x, z) = e^{-t/z} +$$

$$\int_0^x k(|t - y|)J(x - y, x, z)\, dy, \qquad x \geq t. \quad (2)$$

Differentiation of Eq. (2) with respect to x   then gives

$$\frac{d}{dx} J(x - t, x, z) = k(x - y)J(0, x, z) +$$

$$\int_0^x k(|t - y|) \frac{d}{dx} J(x - y, x, z)\, dy. \quad (3)$$

Equation (3) is an integral equation for the function $\frac{d}{dx} J(x - t, x, z)$ whose solutions is given by

$$\frac{d}{dx} J(x - t, x, z) = J(0, x, z)\Phi(t, x),$$

$$= Y(x, z)\int_0^1 J(t, x, z')w(z') \, dz'. \quad (4)$$

Letting $t = 0$ in Eq. (4) gives the desired result

$$\frac{d}{dx} X(x, z) = Y(x, z)\int_0^1 Y(x, z')w(z') \, dz'. \quad (5)$$

The equation for Y and the initial conditions at $x = 0$ are obtained as before.

10. SUMMARY OF THE INITIAL VALUE PROBLEM

At this stage it will be useful to summarize the results obtained so far. We begin with the integral equation for J

$$J(t, x, z) = e^{-(x-t)/z} + \int_0^x k(|t - y|)J(y, x, z) \, dy,$$

$$x \geq t, \qquad 0 \leq z \leq 1. \quad (1)$$

The function J satisfies the Cauchy problem

$$J_x(t, x, z) = -\frac{1}{z} J(t, x, z) +$$

$$X(x, z) \int_0^1 J(t, x, z')w(z') \, dz', \quad (2)$$

with initial condition at $x = t$ given by

$$J(t, t, z) = X(t, z). \qquad (3)$$

We produce the function $X(x, z)$ by means of the differential integral equations

$$X_x(x, z) = Y(x, z) \int_0^1 Y(x, z')w(z') \, dz', \qquad (4)$$

$$Y_x(x, z) = -\frac{1}{z} Y(x, z) +$$

$$X(x, z) \int_0^1 Y(x, z')w(z') \, dz', \quad x > 0. \quad (5)$$

At $x = 0$, the initial conditions for X and Y are given by

$$X(0, z) = Y(0, z) = 1. \qquad (6)$$

As pointed out earlier, we may now compute the function $\Phi(t, x)$ by means of the relation

$$\Phi(t, x) = \int_0^1 J(t, x, z')w(z') \, dz'. \qquad (7)$$

11. SOLUTION PROCEDURE

To make use, numerically, of the equations for X, Y,
and J, the differential-integral equations (10.2), (10.4),
and (10.5) are reduced to an approximate system of ordinary
differential equations. One method that has been useful in
accomplishing this task is to replace the integrals by finite
sums through use of a quadrature scheme. For example, let
$f(x)$ be an integrable function in [0, 1]. Then we write

$$\int_0^1 f(x) \, dx \simeq \sum_{i=1}^{N} f(x_i)\alpha_i, \tag{1}$$

where the numbers $\{x_i\}_{i=1}^{N}$ and $\{\alpha_i\}_{i=1}^{N}$ are appropriate
nodes and weights for the quadrature method in use. For the
case of Gaussian quadrature, x_i is the ith root of the
shifted Legendre polynomial $P_N^*(x)$, which is defined by

$$P_N^*(x) = P_N(1 - 2x), \qquad 0 \leq x \leq 1, \tag{2}$$

where $P_N(x)$ is the Legendre polynomial of degree N given
on $-1 \leq x \leq 1$. The numbers $\{\alpha_i\}$ are then the corresponding
Christoffel weights. Tables of these roots and weights are
readily available in many mathematical handbooks.

The application of these ideas to the equations for
X, Y, and J gives the new system

$$X_i'(x) = Y_i(x) \sum_{j=1}^{N} Y_j(x)w_j, \quad x \geq 0, \tag{3}$$

$$Y_i'(x) = - \frac{1}{z_i} Y_i(x) + X_i(x) \sum_{j=1}^{N} Y_j(x)w_j,$$

$$x \geq 0, \tag{4}$$

$$J_i'(t, x) = - \frac{1}{z_i} J_i(t, x) +$$

$$X_i(x) \sum_{j=1}^{N} J_j(t, x)w_j, \quad x \geq t. \tag{5}$$

where the notation

$$X_i(x) = X(x, z_i), \tag{6}$$

$$Y_i(x) = Y(x, z_i), \tag{7}$$

$$J_i(t, x) = J(t, x, z_i), \tag{8}$$

$$w_i = w(z_i)\alpha_i, \quad i = 1,2,\ldots,N, \tag{9}$$

has been adopted. The initial conditions are

$$X_i(0) = 1, \tag{10}$$

$$Y_i(0) = 1, \tag{11}$$

$$J_i(t, t) = X_i(t), \qquad i = 1, 2, \ldots, N. \tag{12}$$

Equations (3)-(5), together with the initial conditions of Eqs. (10)-(12), comprise the desired set of ordinary differential equations which we shall use for numerical computation.

12.   NUMERICAL EXAMPLES

To illustrate the ideas presented thus far, consider an integral equation which arises in radiative transfer

$$s(t) = \frac{\lambda}{4} e^{-(x-t)/\mu} +$$
$$\frac{\lambda}{2} \int_0^x E_1(|t - t'|)s(t') \, dt', \tag{1}$$

where the kernel is the first exponential integral function

$$E_1(r) = \int_0^1 e^{-r/z'} dz'/z', \qquad r > 0. \tag{2}$$

For the sake of illustration, assume the parameters have the values $\lambda = 0.5$, $x = 0.3$, and $\mu = 0.5$. Then, in terms of our initial value system, the function $w(z)$ for this example is $w(z) = 1/4z$, while the forcing term is $\exp[-2(0.3 - t)]/8$.

The numerical solution was carried out using the

computer program in the appendix. All integrals occurring were evaluated by means of a Gaussian quadrature of order $N = 7$, and a fourth-order Adams-Moulton integration procedure was used to integrate the Cauchy system from $x = 0$ to $x = 0.3$. The integration step size used was $\Delta = 0.005$. Table 1 displays the result of this experiment, along with calculations carried out by Sobolev and Minin for the same problem.

Table 1:

Solutions for Sample Problem with

$\lambda = 0.5$, $x = 0.3$, $\mu = 0.5$.

t	Invariant Imbedding	Sobolev & Minin
0.0	0.082559	0.083
0.1	0.102958	0.103
0.2	0.122851	0.123
0.3	0.140966	0.141

It was shown earlier that the Fredholm resolvent $K(t, y, x)$ also satisfied an initial value problem known as the Bellman-Krein equation. Specifically, the equation is

$$K_x(t, y, x) = \Phi(y, x)\Phi(t, x), \quad x \geq t, \tag{3}$$

with initial condition at $x = y$ given by

$$K(t, y, y) = \Phi(t, y). \tag{4}$$

Use of the equations for X, Y, and J allow calculation of

the resolvent kernel K in the same manner as above. For

example, consider the equation

$$u(t) = g(t) + \frac{1}{2} \int_0^1 e^{-|t-y|} u(y) \, dy, \qquad 0 \le t \le 1. \quad (5)$$

Introduce the variable interval length x. The case

of interest is x = 1. In this example, the parameter z

takes on the single value z = 1. The function J then

satisfies the integral equation

$$J(t, x) = e^{-(x-t)} + \frac{1}{2} \int_0^x e^{-|t-y|} J(y, x) \, dy. \quad (6)$$

The relation between J(t, x) and $\Phi(t, x)$ is $\Phi(t, x) =$
$\frac{1}{2}$ J(t, x). For x = 1, the closed form solution is

$$J(t) = 2/3(1 + t), \qquad 0 \le t \le 1. \quad (7)$$

An integral equation for K when x = 1 is

$$K(t, y) = \frac{1}{2} e^{-|t-y|} +$$

$$\frac{1}{2} \int_0^1 K(t, t') e^{-|t'-t|} \, dt'. \quad (8)$$

Viewing y as a variable, while t is held fixed,

differentiate both sides of Eq. (8) with respect to y. A second differentiation yields $K_{yy} = 0$, so that K is linear in y. However, there is a discontinuity in $K_y(y, t)$ evaluated at y = t,

$$\lim_{y \to t^-} K_y(y, t) - \lim_{y \to t^+} K_y(y, t) = 1. \tag{9}$$

Condition (9), together with the continuity of k at y = t, as well as the knowledge that

$$K(t, 1) = \Phi(t) = \frac{1}{2} J(t) = \frac{1}{3} (1 + t) \tag{10}$$

and

$$K(t, 0) = \Phi(1 - t) = \frac{1}{2} J(1 - t) = \frac{1}{3} (2 - t) \tag{11}$$

give the analytic form of K,

$$K(t, y) = \begin{cases} \dfrac{1}{3} (2 - t) + \dfrac{1}{3} (2 - t)y, & y < t \\[2ex] \dfrac{2}{3} (1 + t) - \dfrac{1}{3} (1 + t)y, & t < y \end{cases} \tag{12}$$

for the interval length x = 1. Equation (12) will serve as a computational check.

First, X and Y are produced by integrating the equations

$$\dot{X}(x) = \frac{1}{2} Y(x)Y(x), \quad X(0) = 1, \tag{13}$$

$$\dot{Y}(x) = -Y(x) + \frac{1}{2} X(x)Y(x), \quad Y(0) = 1, \tag{14}$$

from $x = 0$ to $x = t_1$. The initial conditions at $x = t_1$

$$J(t_1, t_1) = X(t_1), \tag{15}$$

$$K(t_1, t_1, t_1) = \Phi(t_1, t_1) = \frac{1}{2} X(t_1), \tag{16}$$

are imposed, and the equations

$$\dot{J}(t_1, x) = -J(t_1, x) + \frac{1}{2} X(x)J(t_1, x), \tag{17}$$

$$\dot{K}(t_1, t_1, x) = \frac{1}{2} J(t_1, x) - \frac{1}{2} J(t_1, x), \tag{18}$$

are adjoined to the existing set. Then the combined system
is integrated until $x = t_2$. The conditions

$$J(t_2, t_2) = X(t_2), \tag{19}$$

$$K(t_1, t_2, t_2) = \Phi(t_1, t_2) = \frac{1}{2} J(t_1, t_2), \tag{20}$$

$$K(t_2, t_2, t_2) = \Phi(t_2, t_2) = \frac{1}{2} X(t_2), \tag{21}$$

are imposed, and the differential equations

$$\dot{J}(t_2, x) = -J(t_2, x) + \frac{1}{2} X(x)J(t_2, x), \tag{22}$$

$$\dot{K}(t_1, t_2, x) = \frac{1}{2} J(t_1, x) \cdot \frac{1}{2} J(t_2, x), \qquad (23)$$

$$\dot{K}(t_2, t_2, x) = \frac{1}{2} J(t_2, x) \cdot \frac{1}{2} J(t_2, x) \qquad (24)$$

are adjoined. The enlarged system is then integrated to $x = t_3$, and the procedure continues in this fashion. Finally, the entire system of $2 + (2 + 3 + \cdots + N{+}1)$ equations is integrated until $x = 1.0$, the desired length. Only the triangular matrix $K(t_i, t_j, x)$, $J = 1, 2, \ldots, N$, $i = 1, 2, \ldots, j$ is calculated, the remainder of the matrix is obtained by the symmetry relation $K(t_i, t_j, x) = K(t_j, t_i, x)$.

Computational results confirm the effectiveness of the above scheme. The resolvent kernel is calculated to an accuracy of six decimal places using a Gaussian quadrature of order $N = 7$. Computing time on an IBM 7044 is of the order of 5–10 seconds for an integration of 200 steps on the interval $(0, 1)$.

13. VALIDATION OF THE CAUCHY PROBLEM - I

In order to consolidate the analytic and computational gains provided by the initial value procedure just presented and to deepen our understanding of the Cauchy problems, it is important to show, conversely, that the solution of the

Cauchy problem satisfies the original integral equation.

Our approach to the proof of the equivalence between

the initial value problems and the integral equation will be

based upon the uniqueness of the solutions of the Cauchy

problems. Henceforth, we assume the weighting function $w(z)$

is such that the differential-integral equations for X, Y,

and J have unique solutions for all interval lengths

$0 \leq x \leq T$, where T is sufficiently small.

14. OUTLINE OF THE PROOF

Since our proof requires that we follow a somewhat

circuitous route, it is important to lay out the general plan,

lest the unwary reader becomes inextricably entangled in the

plethora of details.

We begin with the initial value problems

$$X_x(x, z) = Y(x, z)\int_0^1 Y(x, z')w(z')\, dz', \qquad (1)$$

$$Y_x(x, z) = -z^{-1}Y(x, t) +$$

$$X(x, z)\int_0^1 Y(x, z')w(z')\, dz', \qquad (2)$$

$$J_x(t, x, z) = -z^{-1}J(t, x, z) +$$

$$X(x, z)\int_0^1 J(t, x, z')w(z') \, dz', \quad (3)$$

$$X(0, z) = Y(0, z) = 1, \quad (4)$$

$$J(t, t, z) = X(t, z). \quad (5)$$

We wish to establish the fact that under the assumption of uniqueness of solution for Eqs. (1)-(3), the function $J(t, x, z)$, produced via the Cauchy problem, satisfies the integral equation

$$J(t, x, z) = e^{-(x-t)/z} +$$

$$\int_0^x k(|t - y|)J(y, x, z) \, dy, \quad (6)$$

where the kernel k is as defined earlier.

At various stages of the proof, the auxiliary functions M, P, W, ψ, ϕ, and Q are introduced. Their definitions are given by the relations

$$M(t, x, z) = e^{-(x-t)/z} +$$

$$\int_0^x k(|t - y|)J(y, x, z) \, dy, \quad 0 \le t \le x, \quad (7)$$

$$P(x, z) = 1 + \int_0^x k(y)J(x - y, x, z) \, dy,$$

$$0 \le x, \quad (8)$$

$$W_x(t, x, z) = Y(x, z)\int_0^1 J(t, x, z')w(z') \, dz',$$

$$x \ge t, \quad (9)$$

$$W(t, t, z) = Y(t, z), \quad (10)$$

$$\psi(t, x, z) = z^{-1}J(t, x, z) +$$

$$Y(x, z)\int_0^1 W(t, x, z')w(z') \, dz' -$$

$$X(x, z)\int_0^1 J(t, x, z')w(z') \, dz', \quad x \ge t, \quad (11)$$

$$\phi(t, x, z) = -z^{-1}W(t, x, z) -$$

$$Y(x, z)\int_0^1 J(t, x, z')w(z') \, dz' +$$

$$X(x, z)\int_0^1 W(t, x, z')w(z') \, dz', \quad x \ge t, \quad (12)$$

$$Q(x, z) = e^{-x/z} + \int_0^x k(y)J(y, x, z) \, dy, \quad x \ge 0. \quad (13)$$

We also define the function Φ as before

$$\Phi(t, x) = \int_0^1 J(t, x, z')w(z') dz', \quad x \geq t. \qquad (14)$$

The following results are established to show the equivalence of the functions J and M.

(i) The functions M and P satisfy the differential equations

$$M_x(t, x, z) = -z^{-1}M(t, x, z) +$$

$$X(x, z)\int_0^1 M(t, x, z')w(z') dz, \qquad (15)$$

$$P_x(x, z) = Y(x, z)\int_0^1 \{e^{-x/z'} +$$

$$\int_0^x k(y)J(y, x, z') dy\}w(z') dz'. \qquad (16)$$

For $t = 0$, we have

$$J(0, x, z) = Y(x, z) \qquad (17)$$

and

$$W(0, x, z) = X(x, z). \qquad (18)$$

(ii) The partial derivatives with respect to t of the functions J and W are

$$J_t(t, x, z) = \psi(t, x, z) \tag{19}$$

$$W_t(t, x, z) = \phi(t, x, z). \tag{20}$$

The total derivative of the function $J(x - t, x, z)$ with respect to x is then given by

$$\frac{d}{dx} J(x - t, x, z) =$$

$$Y(x, z) \int_0^1 W(x - t, x, z')w(z') \, dz'. \tag{21}$$

(iii)   The functions W and J are related by the equation

$$W(x - t, x, z) = J(t, x, z) \tag{22}$$

so that

$$\frac{d}{dx} J(x - t, x, z) = Y(x, z) \int_0^1 J(t, x, z')w(z') \, dz',$$

(iv)   The functions P, Q, X, and Y are connected by the formulas

$$Q(x, z) = Y(x, z), \tag{24}$$

$$P(x, z) = X(x, z).\tag{25}$$

The equivalence of P and X then establishes the desired

result

$$M(t, x, z) = J(t, x, z).\tag{26}$$

15.   THE DIFFERENTIAL EQUATIONS FOR M AND P

Differentiating both sides of Eq. (14.7) with respect

to x   shows that

$$M_x(t, x, z) = -z^{-1}e^{-(x-t)/z} + k(x - t)J(x, x, z) +$$

$$\int_0^x k(|t - y|)J_x(y, x, z)\ dy.\tag{1}$$

According to the differential equation (14.3) for J, the

last integral becomes

$$\int_0^x k(|t - y|)J_x(y, x, z)\ dy =$$

$$\int_0^x k(|t - y|)\{-z^{-1}J(y, x, z) +$$

$$X(x, z)\int_0^1 J(y, x, z')w(z)\ dz'\}\ dy.\tag{2}$$

In view of Eqs. (14.3) and (14.7), the differential equation

(14.15) for the function M is now established. To establis

the initial condition at x = t for the function M, we writ

$$M(t, t, z) = 1 + \int_0^t k(|t - y|)J(y, t, z) \, dy, \qquad (3)$$

or

$$M(x, x, z) = 1 + \int_0^x k(x - y)J(y, x, z) \, dy. \qquad (4)$$

Making the substitution

$$y' = x - y, \qquad (5)$$

the last equation becomes

$$M(x, x, z) = P(x, z) = 1 +$$

$$\int_0^x k(y')J(x - y', x, z) \, dy'. \qquad (6)$$

Differentiating Eq. (6) with respect to x yields

$$P_x(x, z) = k(x)J(0, x, z) +$$

$$\int_0^x k(y) \frac{d}{dx} J(x - y, x, z) \, dy. \qquad (7)$$

Return now to Eq. (14.3) for the function J.

Put t = 0 and observe that the function J(0, x, z)

satisfies the same Cauchy problem that $Y(x, z)$ does, since

$$J(0, 0, z) = X(0, z) = 1. \tag{8}$$

It follows that

$$J(0, x, z) = Y(x, z),$$

which is Eq. (14.17). Similarly, Eq. (14.18) is seen to hold.

16. THE PARTIAL DERIVATIVES OF J AND W

According to Eq. (14.3), the function $J_t = J_t(t, x, z)$ satisfies the differential equation

$$(J_t)_x = -z^{-1} J_t +$$

$$X(x, z) \int_0^1 J_t(t, x, z')w(z')\, dz', \tag{1}$$

Let

$$J_1(t, x, z) = J_t(t, x, z) \tag{2}$$

and

$$J_2(t, x, z) = J_x(t, x, z). \tag{3}$$

Since

$$J(t, t, z) = X(t, z), \tag{4}$$

it follows that at $x = t$

$$J_1(t, t, z) = X_t(t, z) - J_2(t, t, z) =$$

$$Y(t, z)\int_0^1 Y(t, z')w(z') \, dz' + z^{-1}X(t, z) -$$

$$X(t, z)\int_0^1 X(t, z')w(z') \, dz'. \quad (5)$$

The function ψ defined by Eq. (14.11) clearly satisfies the
same initial condition at $x = t$,

$$\psi(t, t, z) = z^{-1}X(t, z) +$$

$$Y(t, z)\int_0^1 Y(t, z')w(z') \, dz' -$$

$$X(t, z)\int_0^1 X(t, z')w(z') \, dz'. \quad (6)$$

In addition, differentiation of Eq. (14.11) with respect to
x   shows that

$$\psi_x(t, x, z) = \frac{1}{z} \, [-z^{-1}J(t, x, z) +$$

$$X(x, z)\int_0^1 J(t, x, z')w(z') \, dz'] + \{-z^{-1}Y(x, z) +$$

$$X(x, z)\int_0^1 Y(x, z')w(z') \, dz\} \int_0^1 W(t, x, z')w(z') \, dz' +$$

$$Y(x, z) \int_0^1 Y(x, z') \int_0^1 J(t, x, z'')w(z'') \, dz'' \, w(z') \, dz' -$$

$$Y(x, z) \int_0^1 Y(x, z')w(z') \, dz' \int_0^1 J(t, x, z'')w(z'') \, dz'' -$$

$$X(x, z) \int_0^1 \{-z'^{-1} J(t, x, z') +$$

$$X(x, z') \int_0^1 J(t, x, z'')w(z'') \, dz''\}w(z') \, dz'. \qquad (7)$$

Through cancellation of like terms and comparison with Eq. (14.11), it is seen that Eq. (7) becomes

$$\psi_x(t, x, z) = -z^{-1}\psi(t, x, z) +$$

$$X(x, z) \int_0^1 \psi(t, x, z')w(z') \, dz', \qquad t \le x. \quad (8)$$

As solutions of the same Cauchy problems, it follows that

$$J_t(t, x, z) = \psi(t, x, z), \qquad 0 \le t \le x,$$

$$0 \le z \le 1. \qquad (9)$$

According to Eqs. (14.9) and (14.10), the Cauchy problem for the function W_t,

$$W_t(t, x, z) = W_1(t, x, z), \qquad (10)$$

is given by

$$(W_t)_x = Y(x, z) \int_0^1 J_t(t, x, z')w(z') \, dz', \quad t \le x,$$

$$W_1(t, t, z) = -z^{-1}Y(t, z) +$$

$$X(t, z) \int_0^1 Y(t, z')w(z') \, dz' \quad (11)$$

$$-Y(t, z) \int_0^1 X(t, z')w(z') \, dz'. \quad (12)$$

The function ϕ, defined in Eq. (14.12), fulfills the same initial condition at $x = t$,

$$\phi(t, t, z) = -z^{-1}Y(t, z) -$$

$$Y(t, z) \int_0^1 X(t, z')w(z') \, dz' +$$

$$X(t, z) \int_0^1 Y(t, z')w(z') \, dz', \quad (13)$$

where use has been made of Eqs. (14.3) and (14.10). In addition, the function ϕ_x, $x \ge t$, is given by the expression

$$\phi_x(t, x, z) = z^{-1}Y(x, z) \int_0^1 J(t, x, z')w(z') \, dz' -$$

$$[z^{-1}Y(x, z) +$$

$$X(x, z) \int_0^1 Y(x, z')w(z') \, dz'] \int_0^1 J(t, x, z')w(z') \, dz' -$$

$$Y(x, z) \int_0^1 [-z'^{-1}J(t, x, z') +$$

$$X(x, z') \int_0^1 J(t, x, z'')w(z'') \, dz'']w(z') \, dz' +$$

$$Y(x, z) \int_0^1 Y(x, z')w(z') \, dz' \int_0^1 W(t, x, z'')w(z'') \, dz'' +$$

$$X(x, z) \int_0^1 w(z') \, dz'Y(x, z') \int_0^1 J(t, x, z'')w(z'') \, dz''.$$

$$(14)$$

Through cancellation of terms and comparison with Eq. (14.11), this expression becomes

$$\phi_x(t, x, z) = Y(x, t) \int_0^1 \psi(t, x, z')w(z') \, dz',$$

$$t \le x, \qquad 0 \le z \le 1. \quad (15)$$

In view of Eq. (9), it follows that the functions ϕ and W_t satisfy the same Cauchy problem so that

$$W_t(t, x, z) = \phi(t, x, z), \qquad t \le x,$$

$$0 \le z \le 1. \quad (16)$$

This establishes Eqs. (14.19) and (14.20).

Keeping Eqs. (14.19), (14.3), and (3) in mind, we see

that

$$\frac{d}{dx} J(x - t, x, z) = \phi(x - t, x, z) +$$

$$J_2(x - t, x, z) = z^{-1}J(x - t, x, z) +$$

$$Y(x, z)\int_0^1 W(x - t, x, z')w(z')\ dz' -$$

$$X(x, z)\int_0^1 J(x - t, x, z')w(z')\ dz' -$$

$$z^{-1}J(x - t, x, z) +$$

$$X(x, z)\int_0^1 J(x - t, x, z')w(z')\ dz' =$$

$$Y(x, z)\int_0^1 W(x - t, x, z')w(z')\ dz'. \qquad (17)$$

This establishes Eq. (14.21).

17. THE RELATION BETWEEN W AND J

To verify Eq. (14.22), we have but to produce the Cauchy problem for the function $W(x - t, x, z)$ and compare it against that for the function J, which is given by Eq. (14.3). For $x = t$ we have

$$W(0, t, z) = X(t, z), \qquad (1)$$

according to Eq. (14.18). Using Eqs. (16.16) and (14.9), it

is seen that

$$\frac{d}{dx} W(x - t, x, z) = -z^{-1}W(x - t, x, z) -$$

$$Y(x, z)\int_0^1 J(x - t, x, z')w(z')dz' +$$

$$X(x, z)\int_0^1 W(x - t, x, z')w(z') dz' +$$

$$Y(x, z)\int_0^1 J(x - t, x, z')w(z') dz', \qquad t \le x. \qquad (2)$$

Upon simplification, the last equation becomes

$$\frac{d}{dx} W(x - t, x, z) = -z^{-1}W(x - t, x, z) +$$

$$X(x, z)\int_0^1 W(x - t, x, z')w(z') dz', \qquad t \le x. \qquad (3)$$

Again assuming uniqueness of solution, it follows that

$$W(x - t, x, z) = J(t, x, z),$$

$$0 \le t \le x \le T$$

$$0 \le z \le 1. \qquad (4)$$

From this representation, it is immediately seen that

$$\frac{d}{dx} J(x - t, x, z) = Y(x, z) \int_0^1 J(t, x, z')w(z') \, dz',$$

$$(5)$$

which is Eq. (14.21).

18. RELATIONS AMONG P, Q, X AND Y

We are now ready to return to Eq. (15.7). It may be written as

$$P_x(x, z) = k(x)Y(x, z) +$$

$$\int_0^x k(y)Y(x, z) \int_0^1 J(y, x, z')w(z') \, dz' \, dy =$$

$$Y(x, z) \int_0^1 \{ e^{-x/z'} +$$

$$\int_0^x k(y)J(y, x, z') \, dy \} w(z') \, dz', \quad (1)$$

where use has been made of the definition of the kernel k. This establishes Eq. (14.16).

The function Q of Eq. (14.13), which is the term in brackets in the last equation, may now be considered. Differentiate both sides of Eq. (14.13) with respect to x to obtain the equation

$$Q_x(x, z) = -z^{-1}e^{-x/z} + k(x)J(x, x, z) +$$

$$\int_0^x k(y) [-z^{-1} J(y, x, z) +$$

$$X(x, z) \int_0^1 J(y, x, z') w(z') \, dz'] \, dy =$$

$$-z^{-1} Q(x, z) + X(x, z) [k(x) +$$

$$\int_0^x k(y) \, dy \int_0^1 J(y, x, z') w(z') \, dz'] =$$

$$-z^{-1} Q(x, z) + X(x, z) \int_0^1 Q(x, z') w(z') \, dz'. \qquad (2)$$

This is Eq. (14.2). Furthermore, at $x = 0$ we have the initial condition

$$Q(0, z) = 1, \qquad 0 \le z \le 1. \qquad (3)$$

By comparing the Cauchy problems for the functions Q and Y, we conclude that

$$Q(x, z) = Y(x, z), \qquad 0 \le x \le T$$
$$0 \le z \le 1. \qquad (4)$$

Equation (1) becomes

$$P_x(x, z) = Y(x, z) \int_0^1 Y(x, z') w(z') \, dz', \qquad (5)$$

and for the initial condition at $x = 0$ we have

$$P(0, z) = 1, \qquad 0 \leq z \leq 1, \tag{6}$$

by the definition of the function P. Comparing this result with Eq. (14.1) for the function X, we see that Eq. (14.25) holds,

$$P(x, z) = X(x, z), \qquad 0 \leq x \leq T$$
$$0 \leq z \leq 1. \tag{7}$$

19.   THE INTEGRAL EQUATIONS FOR J AND Φ

We are now ready to return to the function M. Our earlier discussion showed that it satisfies the same differential equation as the function J. From Eqs. (18.7), (14.8) and (15.6), we see that

$$M(t, t, z) = 1 + \int_0^t k(y) J(x - y, x, z) \, dy$$
$$= P(t, z) = X(t, z). \tag{1}$$

Since the functions J and M satisfy the same Cauchy problems, the uniqueness assumption guarantees that

$$J(t, x, z) = M(t, x, z) \tag{2}$$

or

$$J(t, x, z) = e^{-(x-t)/z} + \int_0^x k(|t - y|) J(y, x, z) \, dy,$$

$$0 \leq t \leq x \leq T, \qquad 0 \leq z \leq 1. \tag{3}$$

This establishes the fundamental result that the function J satisfies the family of Fredholm integral equations in Eq. (14.6).

It is now a simple matter to confirm Eq. (14.14) for the function Φ. By multiplying both sides of Eq. (3) by $w(z)$, integrating on z from 0 to 1, and employing the definition of the functions Φ and k, we see that

$$\Phi(t, x) = k(x - t) + \int_0^1 k(|t - y|)\Phi(y, x) \, dy,$$

$$0 \leq t \leq x \leq T. \tag{4}$$

20.   GENERAL FORCING TERM

In this section we wish to demonstrate that our initial value methods readily extend to the case of a general forcing function $g(t)$.

Let us consider the integral equation

$$u(t) = g(t) + \int_0^c k(|t - y|)u(y) \, dy \tag{1}$$

where g is a continuous forcing function, and k is as defined previously. Imbedding Eq. (1) as before, we obtain the family of equations

$$u(t, x) = g(t) + \int_0^x k(|t - y|)u(y, x) \, dy,$$

$$x \geq t. \quad (2)$$

We shall now obtain a Cauchy problem for the function u.

Differentiate Eq. (2) with respect to x to obtain

$$u_x(t, x) = k(x - t)u(x, x) +$$

$$\int_0^x k(|t - y|)u_x(y, x) \, dy, \quad x \geq t. \quad (3)$$

Viewing Eq. (3) as a linear Fredholm integral equation for the function u_x, we may write its solution as

$$u_x(t, x) = \Phi(t, x)u(x, x), \quad (4)$$

where Φ is as given in Eq. (19.4). Our previous discussion has shown the development of an initial value system for the function Φ; hence, we need now consider only the function $u(x, x)$.

By definition, $u(x, x)$ satisfies

$$u(x, x) = g(x) + \int_0^x k(x - y)u(y, x) \, dy, \quad x \geq 0. \quad (5)$$

Making use of the representation for k as given in Eq.

(2.2), we write Eq. (5) as

$$u(x, x) = g(x) + \int_0^x \int_0^1 e^{-(x-y)/z} w(z) \, dz \, u(y, x) \, dy$$

$$= g(x) + \int_0^1 e(x, z) w(z) \, dz, \qquad (6)$$

where the new function

$$e(x, z) = \int_0^x e^{-(x-y)/z} u(y, x) \, dy \qquad (7)$$

has been introduced and the order of integration interchanged.
To obtain a differential equation for the function e, differentiate Eq. (7) with respect to x obtaining

$$e_x(x, z) = u(x, x) - z^{-1} e(x, z) +$$

$$+ \int_0^x e^{-(x-y)/z} u_x(y, x) \, dy. \qquad (8)$$

For the integral term we may write

$$\int_0^x e^{-(x-y)/z} u_x(y, x) \, dy =$$

$$\int_0^x e^{-(x-y)/z} \phi(y, x) u(x, x) \, dy. \qquad (9)$$

Our last task will be to find a useful expression for the integral

$$I = \int_0^x e^{-(x-y)/z} \Phi(y, x) \, dy. \tag{10}$$

To do this, we use a lemma. Let

$$u_1(t) = g_1(t) + \int_0^x k(|t - y|)u_1(y) \, dy, \tag{11}$$

$$u_2(t) = g_2(t) + \int_0^x k(|t - y|)u_2(y) \, dy. \tag{12}$$

Then by a cross-multiplication and cancellation of like terms, it is readily seen that

$$\int_0^x g_1(y)u_2(y) \, dy = \int_0^x g_2(y)u_1(y) \, dy. \tag{13}$$

Application of this lemma to Eq. (10) yields

$$I = \int_0^x e^{-(x-y)/z} \Phi(y, x) \, dy =$$

$$\int_0^x J(y, x, z)k(x - y) \, dy. \tag{14}$$

Using the integral equation for J as given by Eq. (19.3), we have

$$\int_0^x k(x - y)J(y, x, z) \, dy = J(x, x, z) - 1,$$

$$= X(x, z) - 1. \tag{15}$$

This relation allows the differential equation for the function e to be written as

$$e_x(x, z) = -z^{-1}e(x, z) + X(x, z)u(x, x), \tag{16}$$

or

$$e_x(x, z) = -z^{-1}e(x, z) +$$

$$X(x, z)[g(x) + \int_0^1 e(x, z')w(z') \, dz']. \tag{17}$$

Through Eq. (7) we see that the initial condition for e is given by

$$e(0, z) = 0. \tag{18}$$

21. SUMMARY OF COMPLETE CAUCHY PROBLEM

The complete Cauchy problem for an integral equation with a general forcing term is given by the following system of equations:

$$X_x(x, z) = Y(x, z)\int_0^1 Y(x, z')w(z') \, dz', \quad x > 0, \tag{1}$$

$$Y_x(x, z) = -z^{-1}Y(x, z) +$$

$$X(x, z)\int_0^1 Y(x, z')w(z')\,dz', \qquad x > 0, \quad (2)$$

$$J_x(t, x, z) = -z^{-1}J(t, x, z) +$$

$$X(x, z)\int_0^1 J(t, x, z')w(z')\,dz', \qquad x \geq t, \quad (3)$$

$$e_x(x, z) = -z^{-1}e(x, z) +$$

$$X(x, z)[g(x) + \int_0^1 e(x, z')w(z')\,dz'], \qquad x > 0, \quad (4)$$

$$u_x(t, x) = [g(x) + \int_0^1 e(x, z')w(z')\,dz']$$

$$\int_0^1 J(t, x, z')w(z')\,dz', \qquad x \geq t. \quad (5)$$

The initial conditions are

$$X(0, z) = 1, \tag{6}$$

$$Y(0, z) = 1, \tag{7}$$

$$J(t, t, z) = X(t, z), \tag{8}$$

$$e(0, z) = 0, \tag{9}$$

$$u(t, t) + g(t) + \int_0^1 e(t, z')w(z') \, dz'. \tag{10}$$

The numerical solution procedure follows exactly as outlined in Section 11. As an example, consider the integral equation

$$u(t) = (t - 1) + \frac{1}{2} \int_0^2 E_1(|t - t'|)u(t') \, dt', \tag{11}$$

where again $E_1(r)$ is the first exponential integral function. Earlier investigators give the result $u(2) = 1.520$. Using the computer program in the appendix with a Gaussian quadrature of order $N = 7$ and an integration step size of $\Delta = 0.005$, we obtained the value $u(2) = 1.519455$.

22. VALIDATION OF THE CAUCHY PROBLEM-II

Earlier we proved that for the exponential forcing function, the solutions of the Cauchy problems for X, Y, and J were such that J satisfied the integral equation

$$J(t, x, z) = e^{-(x-t)/z} +$$

$$\int_0^x k(|t - y|)J(y, x, z) \, dy. \tag{1}$$

As a corollary we obtained the result

$$\Phi(t, x) = \int_0^1 J(t, x, z')w(z') \, dz'. \qquad (2)$$

Let us now show that these results together with the Cauchy problems for the functions e and u will establish the validity of our initial value procedure in the case of a general forcing term.

Define the functions A, L, and E by the relations

$$A(t, x) = g(t) + \int_0^x k(|t - y|)u(y, x) \, dy,$$

$$x \geq t, \qquad (3)$$

$$L(t) = \int_0^x k(|t - y|)u(y, t) \, dy, \qquad t > 0, \qquad (4)$$

$$E(x, z) = \int_0^x e^{-(x-y)/z}u(y, x) \, dy,$$

$$x \geq 0, \qquad 0 \leq z \leq 1. \qquad (5)$$

We wish to use the assumption of a unique solution to the Cauchy system to conclude that $A \equiv u$.

To establish this result, differentiate A with respect to x obtaining

$$A_x(t, x) = k(x - t)u(x, x) +$$

$$\int_0^x k(|t - y|)u_x(y, x) \, dy. \tag{6}$$

Making use of Eq. (20.3) for $u_x(t, x)$ and the definition of $u(x, x)$, as well as the integral equation (19.4) for $\Phi(t, x)$, gives

$$A_x(t, x) = [k(x - t) +$$

$$\int_0^x k(|t - y|)\Phi(y, x) \, dy] \times [g(x) +$$

$$\int_0^1 e(x, z')w(z') \, dz']. \tag{7}$$

$$= \Phi(t, x)[g(x) + \int_0^1 e(x, z')w(z') \, dz']. \tag{8}$$

The initial condition for A at $x = t$ is given by Eq. (3) as

$$A(t, t) = g(t) + \int_0^t k(|t - y|)u(y, t) \, dy. \tag{9}$$

We must now show that

$$A(t, t) = g(t) + \int_0^1 e(t, z')w(z') \, dz', \quad t \geq 0. \tag{10}$$

23.   THE INITIAL CONDITION FOR THE FUNCTION A

Let $L(t)$ be defined in the following manner

$$L(t) = \int_0^t k(|t - y|)u(y, t) \, dy. \tag{1}$$

Using the definition of k and interchanging the order of integration yields

$$L(t) = \int_0^1 \left[\int_0^t e^{-(t-y)/z'} u(y, t) \, dy \right] w(z') \, dz'. \tag{2}$$

Now consider the bracketed term in Eq. (2). Define $E(x, z)$ as

$$E(x, z) = \int_0^x e^{-(x-y)/z} u(y, x) \, dy. \tag{3}$$

Differentiate E with respect to x to obtain the formula

$$E_x(x, z) = u(x, x) +$$

$$\int_0^x \{e^{-(x-y)/z}[u_x(y, x) - z^{-1}u(y, x)]\} \, dy. \tag{4}$$

Using Eqs. (3) and (20.3) gives

$$E_x(x, z) = u(x, x)[1 +$$

$$\int_0^x e^{-(x-y)/z} \Phi(y, x) \, dy] - z^{-1} E(x, z). \tag{5}$$

To cope with the bracketed term in Eq. (5), we again appeal to the lemma of Section 20 (Eqs. (20.11)-(20.13)) and the integral equation for J and Φ giving

$$E_x(x, z) = u(x, x)[1 +$$

$$\int_0^x k(x - y) J(y, x, z) \, dy] - z^{-1} E(x, z). \tag{6}$$

Substituting $t = x$ in the integral equation (19.3) for $J(t, x, z)$ shows that

$$E_x(x, z) = -z^{-1} E(x, z) +$$

$$X(x, z)[g(x) + \int_0^1 e(x, z')w(z') \, dz']. \tag{7}$$

The initial condition at $x = 0$ is

$$E(0, z) = 0. \tag{8}$$

This implies that

$$E(x, z) = e(x, z). \tag{9}$$

Using this identity in Eq. (2) now gives

$$L(t) = \int_0^1 e(t, z')w(z') \, dz'. \tag{10}$$

In view of Eqs. (22.9) and (1), this provides the relation

$$A(t, t) = g(t) + \int_0^1 e(t, z')w(z') \, dz'. \tag{11}$$

The initial condition on A thus established, along with Eq. (22.8) and the uniqueness assumption, shows that

$$A(t, x) = u(t, x), \tag{12}$$

or

$$u(t, x) = g(t) + \int_0^x k(|t - y|)u(y, x) \, dy \tag{13}$$

which was to be established.

24. DISCUSSION

We have now shown the complete equivalence between the Fredholm integral equation with displacement kernel of the form (2.2) and the initial value system (21.1)-(21.10).

In the remaining sections of this chapter, we wish to point out how the initial value theory presented above is useful in treating a variety of questions arising in standard treatments of such integral equations. Specifically, we shall discuss the matter of an infinitely long interval of integration, a topic usually associated with the Wiener-Hopf procedure. It will be shown how the imbedding formulation leads to an alternative to the Wiener-Hopf technique. We then turn our attention to the fundamental topic of classical eigenfunctions. We shall demonstrate how these basic functions may be obtained via initial value problems similar to those presented thus far. These results are of extreme importance, since they serve as a bridge between the initial value theory and the vast literature on eigenfunction expansions and function theoretical results for integral operators.

25.   THE HOMOGENEOUS PROBLEM

Our basic tenet to this point has been that invariant imbedding provides a useful tool for the solution of inhomogeneous integral equations. This concentration of attention upon the inhomogeneous problem is in direct contrast to the development of the classical theory of integral equations, in which characteristic values and their associated characteristic elements play the central role. We now wish to investigate the manner in which the invariant imbedding theory may be used to obtain characteristic values and elements.

In order to introduce the basic ideas involved with a minimum of confusion, we first consider the equation

$$J(t, x) = e^{-(x-t)} + \lambda \int_0^x e^{-|t-y|} J(y, x) \, dy, \tag{1}$$

This corresponds to the kernel weighting function being $w(z) = \lambda\delta(z - 1)$, the delta function with jump at $z = 1$. In view of the exponential forcing term, only the initial value problem for the functions X, Y, and J are involved in the solution of Eq. (1). To avoid unnecessary details, assume $\lambda = 1$. Recalling Sec. (10) we see that these equations are

$$X'(x) = Y^2(x), \quad X(0) = 1 \tag{2}$$

$$Y'(x) = -Y(x) + X(x)Y(x), \quad Y(0) = 1, \tag{3}$$

$$J'(t, x) = -J(t, x) +$$
$$X(x)J(t, x), \quad J(t, t) = X(t), \tag{4}$$

where $' = \dfrac{d}{dx}$.

The simple nature of Eqs. (2)-(4) admits the luxury of an explicit analytic solution given by

$$X(x) = 1 + \tan x, \tag{5}$$

$$Y(x) = \sec x, \tag{6}$$

$$J(t, x) = (\sin t + \cos t)\sec x. \tag{7}$$

We now observe that our initial value problem ceases to have a solution at the points $x = (2k + 1)\pi/2$, $k = 0,1,2,\ldots$. The fundamental theorem of analysis which guides us in this situation is the Fredholm Alternative which, roughly speaking, says that the inhomogeneous equation (1) fails to have a unique solution if, and only if, the homogeneous problem

$$\phi(t) = \int_0^x e^{-|t-y|}\phi(y)\ dy \tag{8}$$

has a nontrivial solution. Hence, we suspect that 1 is a characteristic value for the integral operators

$$T_x(\cdot) = \int_0^x e^{-|t-y|}(\cdot)\ dy \tag{9}$$

when $x = (2k + 1)\pi/2$, $k = 0,1,2,\ldots$. We also make the claim that for x sufficiently close to $(2k + 1)\pi/2$, the solution of the inhomogeneous problem, $J(t, x)$, is close to the characteristic element satisfying Eq. (8). We will make these notions precise in a moment.

First, let us note that if our claims are true, then

the invariant imbedding equations contain all the information

necessary to solve both the inhomogeneous and the homogeneous

problems in one sweep (provided, of course, we have a means

to continue the integration beyond the singular lengths).

The procedure, for fixed λ, is to integrate the equations

for the inhomogeneous problem until the solution becomes

large. The solution of the inhomogeneous problem will then

be approximately equal to an unnormalized characteristic

element for the characteristic value λ. We then use one of

the methods to be discussed later to continue the integration

beyond the singular point and proceed as before.

Now let us verify the above claims for our sample

problem. Consider the first critical point $x = \pi/2$. An

elementary calculation gives the Laurent expansion of sec x

and tan x about $x = \pi/2$ as

$$\sec x = \frac{1}{(x-\pi/2)} + \frac{(x-\pi/2)}{6} + \frac{7(x-\pi/2)^3}{60} + \cdots, \qquad (10)$$

$$\tan x = \frac{1}{(x-\pi/2)} - \frac{(x-\pi/2)}{3} - \frac{(x-\pi/2)^3}{45} - \cdots. \qquad (11)$$

Consequently, the functions X, Y, and J may be written as

$$X(x) = \frac{1}{x-\pi/2} + 1 - \frac{(x-\pi/2)}{3} - \frac{(x-\pi/2)^3}{45} - \cdots, \qquad (12)$$

$$Y(x) = \frac{1}{(x-\pi/2)} + \frac{(x-\pi/2)}{6} + \frac{7(x-\pi/2)^3}{60} + \cdots, \qquad (13)$$

and

$$J(t, x) = (\sin t + \cos t)\left[\frac{1}{(x-\pi/2)} + \right.$$

$$\left. \frac{(x-\pi/2)}{6} + \frac{7(x-\pi/2)^3}{60} + \cdots\right]. \qquad (14)$$

Similar expansions obtain for the other singular lengths. Eqs. (12)-(14) show that the functions X, Y, and J all have simple poles at $x = \pi/2$. If we multiply both sides of Eq. (14) by $(x - \pi/2)$, we see that

$$\lim_{x \to \pi/2} (x - \pi/2)J(t, x) = \sin t + \cos t. \qquad (15)$$

It is a simple exercise to verify that the function $\phi(t) = \sin t + \cos t$ is a nontrivial solution of Eq. (8). Thus, our earlier claims are justified for this particular example. It is interesting to note that if $(x - \pi/2) = \varepsilon > 0$, then we have the estimate

$$J(t, x) - \frac{\phi(t)}{\varepsilon} = \frac{\varepsilon}{6} + \frac{7\varepsilon^2}{60} + o(\varepsilon^3) \qquad (16)$$

For example, if $\varepsilon = 10^{-3}$ then Eq. (16) gives

$$J(t, x) - 10^3 \phi(t) \approx 10^{-4}. \qquad (17)$$

The preceding example is typical of the sort of behavior that the initial value problem will possess for the types of integral operators which we are considering in this chapter. In general, the solution of the inhomogeneous problem

$$u(t) = g(t) + \lambda \int_0^x k(|t - y|)u(y) \, dy \qquad (18)$$

has the expansion

$$u(t, x) = \sum_{n=-k}^{\infty} a_n(t)(x - x_i)^n, \qquad 0 \leq x_i \qquad (19)$$

valid for $|x - x_i|$ sufficiently small. (Note that if any $a_n(t)$, $-k \leq n \leq -1$ is not zero, then x_i is a singular length for the value λ.) Knowledge of the order of the pole which $u(t, x)$ has at $x = x_i$, then allows us to assert that

$$u(t, x) \sim \phi(t), \quad |x - x_i| \ll 1, \qquad (20)$$

where $\phi(t)$ is the eigenfunction associated with the number λ for the interval length x_i.

26. CONTINUATION BEYOND SINGULAR LENGTHS

In the previous section we presented a procedure for

obtaining characteristic values and elements from the
invariant imbedding equations under the assumption that we
have means available for continuing the integration of our
initial value problem beyond any singular points. Let us
now take up a more detailed study of how this may be carried
out.

Consider the sample problem

$$J(t, x) = e^{-(x-t)} + \int_0^x e^{-|t-y|} J(y, x) \, dy. \tag{1}$$

We saw earlier that the functions X, Y, and J were given
by

$$X(x) = 1 + \tan x, \tag{2}$$

$$Y(x) = \sec x, \tag{3}$$

$$J(t, x) = (\sin t + \cos t)\sec x. \tag{4}$$

One obvious method which may be useful in continuing
the integration is to change variables when the functions
X, Y, and J become large. For example, the new variables

$$R(x) = 1/X(x), \tag{5}$$

$$S(x) = 1/Y(x), \tag{6}$$

$$T(t, x) = 1/J(t, x), \tag{7}$$

will be passing through zeros when X, Y, and J are
becoming unbounded at $x = (2k + 1)\pi/2$, $k = 0,1,\ldots$. Thus,
when x nears any of the singular points, we shift over to
the new equations

$$R'(x) = -R^2(x)/S^2(x), \tag{8}$$

$$S'(x) = S(x) - S(x)/R(x), \tag{9}$$

and

$$T'(t, x) = -T(t, x) - T(t, x)/R(x), \tag{10}$$

with the initial conditions at $x = x^*$,

$$R(x^*) = 1/X(x^*), \tag{11}$$

$$S(x^*) = 1/Y(x^*), \tag{12}$$

$$T(t, x^*) = 1/J(t, x^*), \tag{13}$$

where x^* is a point such that $(2k + 1)\pi/2 < x^* < (2k + 3)\pi/2$, $k = 0,1,\ldots$. Once the equations for R, S,
and T have been integrated beyond the singular point, we
may then shift back to the original variables. Making use of
the explicit solutions given above, we have for this example

$$R(x) = \frac{\cos x}{\sin x + \cos x} \, , \tag{14}$$

$$S(x) = \cos x, \tag{15}$$

$$T(t, x) = \frac{\cos x}{\cos t + \sin t} \, . \tag{16}$$

Another continuation method is to make use of the Laurent expansions of the functions X, Y, and J about the singular point in question. For the sample problem, we have seen that for x near $\pi/2$

$$X(x) = \frac{1}{(x-\pi/2)} + 1 - \frac{(x-\pi/2)}{3} - \frac{(x-\pi/2)^3}{45} \ldots , \tag{17}$$

$$Y(x) = \frac{1}{(x-\pi/2)} + \frac{(x-\pi/2)}{6} + \frac{7(x-\pi/2)^3}{60} + \cdots , \tag{18}$$

$$J(t, x) = (\sin t + \cos t) \left[\frac{1}{(x-\pi/2)} + \right.$$

$$\left. \frac{(x-\pi/2)}{6} + \frac{7(x-\pi/2)^3}{60} \right] . \tag{19}$$

Thus, to obtain initial values at a point $\pi/2 < x^* < \frac{3\pi}{2}$, we need only evaluate $X(x^*)$, $Y(x^*)$, and $J(t, x^*)$ by means of Eqs. (17)-(19). In general, we would have to evaluate the expansion coefficients by means of an interpolation based upon values of the functions X, Y, and J near the singularity.

For example, the functions X, Y, and J may be evaluated

at the points x_1, x_2, x_3 close to the singular value of x.

Then, assuming a simple pole at the critical point, the

coefficients a_{-1}, a_0, a_1 in the Laurent expansion may be

obtained by a least-squares fit.

Use of the original integral equation, together with

the method of successive approximations, furnishes another

means to continue beyond critical lengths. For example, in

Eq. (1), use of successive approximations would yield the

iterative scheme

$$J_n(t,\ x) = e^{-(x-t)} + \int_0^x e^{-|t-y|} J_{n-1}(y,\ x)\ dy,$$

$$x > \pi/2, \quad n = 1,2,\ldots, \quad (20)$$

$$J_0(t,\ x) = e^{-(x-t)}. \quad (21)$$

It is well known that the sequence $\{J_n(t,\ x)\}$ will diverge

for x larger than the first critical length unless special

precautions are taken. In some cases, use of a Shanks trans-

form on the sequence $\{J_n\}$ has been observed to induce con-

vergence for interval lengths beyond the first critical

length. References to this technique are given at the end of

the chapter. Once the function J(t, x) has been determined,

the values for X and Y are obtained from J by letting

$t = 0$ and x, respectively.

The reduction to linear algebraic equations is another standard solution method for integral equations which can be useful in circumventing singular points. As applied to the sample problem, we would use a quadrature formula of some type to replace the integral in Eq. (1) by a finite sum

$$\int_0^x e^{-|t-y|} J(y,\ x)\ dy = \sum_{i=1}^N \exp(-|t-y_i|) J(y_i,\ x) w_i,$$

$$x > \pi/2. \quad (22)$$

The numbers y_i and w_i are determined by the quadrature scheme employed. The original equation for J now becomes

$$J(t,\ x) = e^{-(x-t)} + \sum_{i=1}^N \exp(-|t-y_i|) J(y_i,\ x) w_i. \quad (23)$$

Letting t assume the values $\{y_i\}$, we see that Eq. (23) reduces to a system of linear algebraic equations for the unknowns $J(y_i,\ x)$. Resolution of the system (23) then gives starting values for J (and hence for X and Y) for interval lengths larger than $\pi/2$.

There is no difficulty in extending the ideas presented here for the simple example of Eq. (1) to more general equations of the type dealt with earlier. The only additional steps necessary are the introduction of more functions to account for the functions e and u which

form a part of the general Cauchy system.

27. INFINITE INTERVALS-I

Many problems in physics and engineering require the

solution of the integral equation

$$u(t) = g(t) + \int_0^\infty k(|t - y|)u(y) \, dy. \qquad (1)$$

For example, Milne's equation of radiative equilibrium in

stellar atmospheres has this form. The classical solution

procedure is to take the Fourier transform of both sides of

Eq. (1) and use the Weiner-Hopf technique to obtain the

unknown function u. As we indicate below, the procedures

that we have thus far developed may be used to furnish an

alternative approach to the Weiner-Hopf method. Keeping the

idea of a feasible computational algorithm foremost in our

minds, we show that the combination of invariant imbedding

and a bit of analytic sleight-of-hand leads to several

attractive possibilities for the solution of Eq. (1).

In view of the fact that we have presented a means

for solving the integral equation

$$u(t, x) = g(t) + \int_0^x k(|t - y|)u(y, x) \, dy \qquad (2)$$

for any finite x for which the solution exists, our first

approach to the solution of Eq. (1) is to consider it as the

limiting form of Eq. (2) as $x \to \infty$. Hence, it is reasonable

to suspect that as $x \to \infty$, the sequence $u(t, x)$ will con-

verge to the solution $u(t)$ of Eq. (1) as $x \to \infty$. Thus, we

may integrate our initial value problem starting at x = 0,

and continue until the solution approaches a limiting value.

Under suitable hypotheses on the functions g and k, this

limiting value will solve Eq. (1).

As simple as this procedure seems, it suffers from

the defect that it may be necessary to carry out the integra-

tion to a very large interval before a limiting behavior is

approached. Hence, too much computation may be required. In

this case, the use of an extrapolation procedure may be

helpful. For example, assume that for a fixed t, the solu-

tion of Eq. (2) for a finite interval x is

$$u(t, x) = A + Be^{-\beta x} \qquad (3)$$

where the number A is the limiting value,

$$A = \lim_{x \to \infty} u(t, x). \qquad (4)$$

Since there are three unknown constants appearing in Eq.

(3), evaluation of $u(t, x)$ at three interval lengths

x_{n-1}, x_n, x_{n+1}, allows the calculation of A by

$$A = \frac{u(t,x_{n+1})u(t,x_{n-1}) - u^2(t,x_n)}{u(t,x_{n+1}) + u(t,x_{n-1}) - 2u(t,x_n)} . \tag{5}$$

Thus, by calculating the solution for three interval lengths

x_n, x_{n+1}, x_{n-1}, we may predict the asymptotic value of

$u(t, x)$ using Eq. (5). This prediction is a special case

of a general class of nonlinear extrapolation formulas.

References to the papers of Shanks and others, where these

ideas are taken up in detail, are given at the end of the

chapter.

28.   INFINITE INTERVALS-II

Let us now make use of the auxiliary functions

introduced in the validation of the Cauchy system for X, Y,

and J to obtain another method which is suitable for the

solution of equations on the half-line in some cases.

Suppose we have the equation

$$u(t) = g(t) + \int_0^\infty k(|t - y|)u(y) \, dy \tag{1}$$

where we now assume that the forcing term $g(t)$ may be
written as

$$g(t) = \int_0^1 e^{-t/z} \alpha(z) \, dz \tag{2}$$

for an appropriate weighting function α. Then we may write
the solution of Eq. (1) as

$$u(t) = \int_0^1 J(t, z) \alpha(z) \, dz, \tag{3}$$

where $J(t, z)$ is the solution of the integral equation

$$J(t, z) = e^{-t/z} + \int_0^\infty k(|t - y|) J(y, z) \, dy,$$

$$t \geq 0, \qquad 0 \leq z \leq 1. \tag{4}$$

We now recall the function $W(t, x, z)$ which was
introduced in Sec. 14. We note that W satisfies the
integral equation

$$W(t, x, z) = e^{-t/z} + \int_0^x k(|t - y|) W(y, x, z) \, dy,$$

$$x \geq t, \qquad 0 \leq z \leq 1. \tag{5}$$

Consequently,

$$\lim_{x \to \infty} W(t, x, z) = J(t, z). \tag{6}$$

Earlier we noted that W and J satisfied the initial value problem

$$W_t(t, x, z) = -z^{-1}W(t, x, z) -$$

$$Y(x, z) \int_0^1 J(t, x, z')w(z') \, dz' +$$

$$X(x, z) \int_0^1 W(t, x, z')w(z') \, dz',$$

$$x \geq t, \quad 0 \leq z \leq 1. \tag{7}$$

$$J_t(t, x, z) = z^{-1}J(t, x, z) +$$

$$Y(x, z) \int_0^1 W(t, x, z')w(z') \, dz' +$$

$$X(x, z) \int_0^1 J(t, x, z')w(z') \, dz', \tag{8}$$

$$W(0, x, z) = X(x, z), \tag{9}$$

$$J(0, x, z) = Y(x, z). \tag{10}$$

Assuming a prior calculation has produced the functions X and Y, we now have another initial value problem for J and W, where t is now the independent variable. The flexibility of having this additional set of equations will become apparent in what follows.

Under the assumption (as is the case in many problems of interest) that $Y(x, z) \to 0$ as $x \to \infty$, we have

$$J_t(t, z) = \lim_{x \to \infty} W_t(t, x, z) \tag{11}$$

$$J_t(t, z) = -z^{-1} J(t, z) +$$

$$\phi(z) \int_0^1 J(t, z') w(z') \, dz', \tag{12}$$

where the function $\phi(z)$ is defined by the relation

$$\phi(z) = \lim_{x \to \infty} X(x, z). \tag{13}$$

To obtain values of $\phi(z)$, we may either perform a preliminary calculation in which the Cauchy problem for X and Y is integrated for sufficiently large x to guarantee a close approximation to the limiting value, or we may make use of the fact that ϕ satisfies the nonlinear integral equation

$$\phi(z) = 1 + \int_0^1 \frac{zv}{z+v} \phi(z)\phi(v)w(v) \, dv,$$

$$0 \le z \le 1. \tag{14}$$

To establish this result, consider the function

$R(v, z, x)$ given by

$$R(v, z, x) = \int_0^x e^{-(x-y)/z} J(y, x, v) \, dy. \tag{15}$$

Differentiating Eq. (15) with respect to x yields

$$R_x(v, z, x) = J(x, x, v) +$$

$$\int_0^x e^{-(x-y)/z} J_x(y, x, v) \, dy -$$

$$\frac{1}{z} \int_0^x e^{-(x-y)/z} J(y, x, v) \, dy. \tag{16}$$

$$= J(x, x, v) - \frac{1}{z} R(v, z, x) +$$

$$\int_0^x e^{-(x-y)/z} \{-v^{-1} J(y, x, v) +$$

$$X(x, v) \int_0^1 J(y, x, v') w(v') \, dv'\} \, dy$$

$$= J(x, x, v) - \left(\frac{1}{z} + \frac{1}{v}\right) R(v, z, x) \tag{17}$$

$$+ X(x, v) \int_0^x \int_0^1 e^{-(x-y)/z} J(y, x, v') w(v') \, dv' \, dy \tag{18}$$

To obtain Eq. (18), use has been made of our previous initial value problem for $J(t, x, v)$. Putting in the definition of $J(x, x, v)$ and interchanging the order of integration in the last term, Eq. (18) becomes

$$R_x(v, z, x) = - \left(\frac{1}{z} + \frac{1}{v}\right) R(v, z, x) +$$

$$X(x, v)[1 + \int_0^1 R(v', z, x)w(v') \, dv'] \qquad (19)$$

Recalling that

$$X(x, z) = 1 + \int_0^x k(x - y)J(y, x, z) \, dy$$

$$= 1 + \int_0^x \int_0^1 e^{-(x-y)/z'} w(z') \, dz' \, J(y, x, z) \, dy, \qquad (20)$$

we see that

$$X(x, z) = 1 + \int_0^1 R(v', z, x)w(v') \, dv'. \qquad (21)$$

Hence, Eq. (19) assumes its final form

$$R_x(v, z, x) = - \left(\frac{1}{z} + \frac{1}{v}\right) R(v, z, x) +$$

$$X(x, v)X(x, z). \qquad (22)$$

The initial condition follows immediately from Eq. (15),

$$R(v, z, 0) = 0. \qquad (23)$$

Observe also the symmetry of R in its first two arguments;

i.e.

$$R(v, z, x) = R(z, v, x). \tag{24}$$

Integration of Eq. (22) yields

$$R(x, z, x) =$$

$$e^{-(1/z+1/v)x} \int_0^x e^{(1/z+1/v)y} X(y, z)X(y, v) \, dy. \tag{25}$$

Substitution from Eq. (25) into Eq. (21) gives the equation

$$X(x, z) = 1 +$$

$$\int_0^1 \{e^{-(1/z+1/v)x} \int_0^x e^{(1/z+1/v)y} X(y, z)X(y, v)\} w(v) \, dv. \tag{26}$$

Letting $x \to \infty$ and assuming that the order of integration and limit operation may be interchanged, we have

$$\phi(z) = 1 +$$

$$\int_0^1 \{\lim_{x \to \infty} e^{-(1/z+1/v)x} \int_0^x e^{(1/z+1/v)y} X(y,z)X(y,v)\} dy \, w(v) dv. \tag{27}$$

L'Hospital's rule provides

$$\phi(z) = 1 + \int_0^1 \lim_{x \to \infty} \frac{X(x,z)X(x,v)}{1/z+1/v} w(v) \, dv \tag{28}$$

or

$$\phi(z) = 1 + \int_0^1 zv \; \frac{\phi(z)\phi(v)}{z+v} \; w(v) \; dv, \tag{29}$$

which was to be established.

The method of successive approximations may now be employed to obtain values of ϕ from Eq. (29). The initial condition to be used for Eq. (8) is

$$J(0, z) = \lim_{x \to \infty} X(x, z) = \phi(z). \tag{30}$$

29. INFINITE INTERVALS - AN EXAMPLE

As an example of this technique, we consider the integral equation of radiative transfer

$$J(t, z) = \frac{\lambda}{4} e^{-t/z} +$$

$$\frac{\lambda}{2} \int_0^\infty E_1(|t - y|)J(y, z) \; dy, \tag{1}$$

where

$$E_1(|t - y|) = \int_0^1 e^{-|t-y|/z} \; dz/z. \tag{2}$$

The initial value problem satisfied by the J function of radiative transfer is

$$J_t(t, z) = -z^{-1}J(t, z) + \frac{\lambda}{4} \phi(z)\Phi(t), \quad t > 0, \qquad (3)$$

$$J(0, z) = \frac{\lambda}{4} \phi(z), \qquad (4)$$

where

$$\Phi(t) = 2 \int_0^1 J(t, z') \, dz'/z'. \qquad (5)$$

Note that the J, ϕ, and Φ function of radiative transfer differ slightly from their analogs above.

Replacing the integrals by gaussian quadratures as before, we obtain a system of ordinary differential equations for J. The function $\phi(z)$ is determined from the nonlinear integral equation

$$\phi(z) = 1 + \lambda/2 \, z\phi(z) \int_0^1 \frac{z'\phi(z')}{z+z'} \, dz', \quad 0 \le z \le 1. \quad (6)$$

To solve this equation, we use an extension of Newton's method. The integral is replaced by a sum according to the N-point Guassian quadrature formula. The resultant equation is linearized. Successive approximations to the N unknowns $\phi(z_1)$, $\phi(z_2)$,..., $\phi(z_N)$ are obtained by solving linear algebraic equations.

As a numerical experiment, we produce the functions ϕ, J, and Φ for $\lambda = 1.0$. A quadrature of order $N = 7$ is used. The function ϕ was evaluated at the 7 nodes of the quadrature formula and was produced to an accuracy of one part in 10^5 in 14 iterations, using an initial approximation $\phi(z) \equiv 1$. The values of z_i and $\phi(z_i)$ are displayed in Table 1. These values of ϕ are used to compute the physically meaningful functions Φ and J, which are given in Table 2.

Checks against previously published results indicate that the calculations are accurate. Note that in Table 2 the functions approach a limiting value with increasing t. The limiting value of Φ according to theory is $\sqrt{3} \cong 1.732$.

Table 1:

The Function $\phi(z)$

z_i	$\phi(z_i)$
.0254	1.0759
.1292	1.3084
.2971	1.6368
.5000	2.0124
.7029	2.3786
.8708	2.6779
.9746	2.8618

Table 2:

Value of J_i and Φ

t	0.2	0.6	1.0	3.0
Φ	2.0223	1.8149	1.7662	1.7313
$J(t,z_1)$	0.0142	0.0125	0.0121	0.0119
$J(t,z_4)$	0.5296	0.4979	0.4707	0.4368
$J(t,z_7)$	0.8850	1.0290	1.1014	1.1965

CHAPTER FOUR

NOTES AND REFERENCES

§1.   Classical treatment of integral equations may be

found in

R. Courant–D. Hilbert, Methods of Mathematical

Physics, Vol. 1, Interscience, New York, 1953.

F. Tricomi, Integral Equations, Interscience, New

York, 1957.

S. Mikhlin, Integral Equations, Pergamon Press, New

York, 1957.

§4.   These results first appeared in

H. Kagiwada, R. Kalaba, and A. Schumitzky, "A

Representation for the Solution of Fredholm Integral

Equations," Proc. Am. Math. Soc., 23, 1969, pp. 37–

40.

§5.   The fundamental equation for the resolvent kernel was

obtained independently by Bellman and Krein in

R. Bellman, "Functional Equations in the Theory of

Dynamic Programming–VII: A Partial Differential

Equation for the Fredholm Resolvent," Proc. Amer.

Math. Soc., 8, 1957, pp. 435–440.

M. Krein, "On a New Method for Solving Linear
Integral Equations of the First and Second Kinds,"
Dokl. Akad. Nauk SSSR, 100, 1955, pp. 413-416.

§6.   The function Φ is of considerable interest in
studies of radiative transfer in the atmosphere.
For further details see

V. V. Sobolev, A Treatise on Radiative Transfer,
Van Nostrand, Princeton, 1963.

§8.   The equations for the functions X and Y are the
same as those given by Chandrasekhar. For their use
in radiative transfer processes see

S. Chandrasekhar, Radiative Transfer, Dover, New
York, 1960.

§10.   The derivation given here follows that in

J. Casti, H. Kagiwada, and R. Kalaba, "The Invariant Imbedding Numerical Method for the Solution of
Fredholm Integral Equations with Displacement
Kernels," Proc. ACM Nat'l. Conf., San Francisco,
1969.

§12.   For further details of the calculations presented
here see

H. Kagiwada and R. Kalaba, "An Initial Value Method Suitable for the Computation of Certain Fredholm Resolvents," <u>J. Math and Phy. Sciences</u>, 1-2, 1967, pp. 109-122.

§13-24. This proof was first published in

J. Casti and R. Kalaba, "Proof of the Basic Invariant Imbedding Method for Fredholm Integral Equations with Displacement Kernels-I," <u>Info. Sciences</u>, 2, 1970, pp. 51-67.

§25.   A graph of the first characteristic value versus singular length for the kernal $E_1(|t - y|)$ is given in

R. Bellman, H. Kagiwada, and R. Kalaba, "New Derivation of the Integro-Differential Equations for Chandrasekhar's X and Y Functions," <u>J. Math. Phys.</u>, 9, 1968, pp. 906-908.

§26.   These results are taken from

J. Casti, R. Kalaba, and M. Scott, "A Proposal for the Calculation of Characteristic Values and Functions for Certain Differential and Integral Operators Via Initial Value Procedures," <u>J. Math. Anal. Applic.</u>, Jan. 1973.

§27-30. For further references to the problem on the half-
line see

H. Kagiwada, R. Kalaba and B. Vereeke, "Invariant
Imbedding and Fredholm Integral Equations with
Displacement Kernels on an Infinite Interval,"
The RAND Corp., RM-5646-PR, May 1968.

Details on nonlinear summability methods and extrapolation to the limit are found in

Shanks, D., "Nonlinear Transformations of Divergent
and Slowly Convergent Sequences," J. Math. and
Physics, 34, 1955/56, 1-42.

Bellman, R. and R. Kalaba, "A Note on Nonlinear
Summability Techniques in Invariant Imbedding,"
J. Math. Anal. Appl., 6, 1963, 465-472.

CHAPTER FIVE

VARIATIONAL PROBLEMS

1. INTRODUCTION

The post-Sputnik era has witnessed major strides forward in the computational treatment of broad classes of optimal guidance and control process. Powerful theories and techniques such as dynamic programming, the gradient method, and the maximum principle have been developed to deal with the thorny analytical and computational difficulties that arise when we are forced to jump from textbook examples to the real-world. The primary motivation for these techniques has been computational, since the traditional formulation of optimal control problems leads to Euler equations and two-point boundary value problems. As usual, the numerical solutions of these boundary value problems is far from routine.

The computational complexities of dealing with

boundary value problems has provided the stimulus to attempt

to characterize the optimizing functions as solutions of

initial value problems. In this paper, we wish to use the

theory of invariant imbedding to indicate the path by which

the solutions of large classes of linear and nonlinear opti-

mization and control problems may be transformed into the

solutions of Cauchy problems.

To establish our basic theme in the simplest possible

setting, we first take up a discussion of the quadratic varia-

tional problem. The Cauchy problem obtained leads to a one-

sweep integration procedure for determining the optimizing

function. Having cut our teeth on the quadratic problem, we

then investigate in succession, nonlinear variational prob-

lems, unconstrained nonlinear control problems, and finally

constrained control processes. At no point during the dis-

cussion is any use made of Euler equations, the Principle of

Optimality, or the Pontryagin Maximum Principle.

2. THE QUADRATIC COST PROBLEM

Consider the problem of minimizing the quadratic

functional

$$I(W) = \int_a^T [W^2 + g(y)W^2(y)]\, dy. \tag{1}$$

subject to

$$W(a) = 1, \tag{2}$$

$$W(T) = \text{free}, \quad T \quad \text{fixed.} \tag{3}$$

The optimizing function clearly depends upon both y and a,

$$W_{opt} = u(y, a), \qquad a \le y \le T. \tag{4}$$

In the usual manner, a small variation about the optimal

solution is introduced,

$$W = u + \varepsilon\eta, \tag{5}$$

where ε is a small parameter, and η is an arbitrary

function of y with the exception that at y = a,

$$\eta\big|_{y=a} = 0. \tag{6}$$

Substituting (5) into (1) and expanding I in powers of ε,

it is readily concluded by the usual arguments that the

coefficient of ε in the expansion must be zero. That is

to say,

$$\int_a^T [\ddot{u}\eta + gu\eta] \, dy = 0, \tag{7}$$

where

$$(\cdot) = \frac{d}{dy} .$$

Employment of integration by parts and application

of the fundamental lemma of the calculus of variations to

Eq. (7) would lead to the Euler equation

$$\ddot{u} - g(y)u = 0, \tag{8}$$

with the associated boundary conditions

$$u(a) = 1, \quad \dot{u}(T) = 0. \tag{9}$$

We wish to use an entirely different approach; it results in a Cauchy problem for determining the optimal arcs.

3.   THE INITIAL VALUE SYSTEM

In Eq. (2.7), rather than integrating by parts, let us make use of the fact that the variation function η is arbitrary by choosing a special function convenient for our purposes. Such a function is

$$\eta = k(t, \ y, \ a), \qquad a \le t \le T, \tag{1}$$

where

$$k = \begin{cases} y - a, & a \le y \le t \\ t - a, & t \le y \le T, \end{cases} \tag{2}$$

or

$$k = \min(y - a, \ t - a). \tag{3}$$

Eq. (2.7) then becomes

$$u(t, \ a) = u(a, \ a) - \int_a^T g(y)k(t, \ y, \ a)u(y, \ a) \ dy,$$

$$a \le t \le T. \tag{4}$$

In view of condition (2.2) on optimal arcs, we see that this is a linear Fredholm integral equation of the second kind for the optimizing function u. A unique solution exists if T - a is sufficiently small, write

$$u(t, a) = 1 - \int_a^T g(y)k(t, y, a)u(y, a) \, dy \qquad (5)$$

Differentiation of both sides of Eq. (5) with respect to a yields the result

$$u_a(t, a) = g(a)k(t, a, a)u(a, a) -$$

$$\int_a^T g(y)k_a(t, y, a)u(y, a) \, dy -$$

$$\int_a^T g(y)k(t, y, a)u_a(y, a) \, dy. \qquad (6)$$

Use of Eqs. (2) and (2.2) allows us to simplify this to

$$u_a(t, a) = \int_a^T g(y)u(y, a) \, dy -$$

$$\int_a^T g(y)k(t, y, a)u_a(y, a) \, dy, \qquad a \le t \le T, \qquad (7)$$

a linear Fredholm integral equation for the function u_a.

Introducing the auxiliary function r as

$$r(a) = \int_a^T g(y)u(y, a) \, dy, \quad a \leq T, \tag{8}$$

and comparing Eqs. (5) and (7), we see that

$$u_a(t, a) = r(a)u(t, a), \quad a < t. \tag{9}$$

This is an ordinary differential equation for the optimizing u, with initial condition given by Eq. (2.2) as

$$u(t, t) = 1. \tag{10}$$

In order to complete our Cauchy system, we must now determine an equation satisfied by r.

Upon differentiating Eq. (8) with respect to a, we obtain

$$r'(a) = -g(a)u(a, a) + \int_a^T g(y)u_a(y, a) \, dy. \tag{11}$$

Eqs. (2.2) and (9) allow the simplification

$$r'(a) = -g(a) + r(a)\int_a^T g(y)u(y, a) \, dy \tag{12}$$

or

$$r'(a) = -g(a) + r^2(a), \quad a < T. \tag{13}$$

The initial condition at a = T is given by Eq. (8) as

$$r(T) = 0. \tag{14}$$

Eqs. (9), (10), (13), and (14) comprise the complete Cauchy

system for determining the optimizing function u.   It should

also be noted that the function $r(a)$ is nothing more than

$-u(a, a)$, the negative of the optimal slope at one end of the

process. Consequently, if all that is desired is this

quantity, only Eqs. (13) and (14) need be integrated.

4. COMPUTATIONAL SCHEME

     The Cauchy system of Section 3 is extremely easy to

use for numerical calculation of u since the boundary value

aspects of the Euler equation have been displaced by an

initial value problem.

     Let us assume that we desire the solution for a pro-

cess of length $T - a = \alpha > 0$. The solution procedure is to

integrate Eq. (3.13) for r subject to the known initial

condition (3.14) from $a = T$ to $a = t_1$, a point where the

function u is desired. At $a = t_1$, adjoin Eq. (3.9) for

the function $u(t_1, a)$ subject to the initial condition

$u(t_1, t_1) = 1$. Then integrate the entire system consisting

of the functions $r(a)$ and $u(t_1, a)$ from $a = t_1$ to $a =$

$T-\alpha$. The function value $u(t_1, T-\alpha)$ will then be the optimal

solution at the point t_1 for a process of duration α. If

the solution is desired at additional points t_2, t_3...
contained in the interval $[\alpha, T]$, additional equations for
$u(t_i, a)$ are adjoined at the appropriate points.

Notice that the foregoing approach solves not only
the original problem of interest, but also all "nearby"
problems. This type of sensitivity analysis is valuable in
many applications.

5.   SIMPLE EXAMPLES

To underscore the fact that the above Cauchy problem
is completely equivalent to the classical formulation, consider the special case of (2.1) in which $a = 0$ and $g = 1$.
For this case, the Cauchy system has the explicit solution

$$r(a) = -\tanh(T - a), \tag{1}$$

$$u(t, a) = \frac{\cosh(T - t)}{\cosh(T - a)}, \tag{2}$$

which for $a = 0$ reduces to

$$u(t, 0) = \frac{\cosh(T - t)}{\cosh T}. \tag{3}$$

This is the same result as one obtains by solving the Euler
equation associated with the given functional.

Now consider the situation in which $g = -1$. The

Riccati equation for the auxiliary function   r   is now

$$r'(a) = 1 + r^2, \quad a < T, \tag{4}$$

$$r(T) = 0. \tag{5}$$

The analytical solution is

$$r(a) = -\tan(T - a). \tag{6}$$

Thus, the function   r   is seen to have a singularity for
$T - a = \frac{\pi}{2}$, $3\pi/2,\ldots$.   In fact, there is a conjugate point
at each of these interval lengths and Eq. (3.13) for   r   can
be recognized as Legendre's form of Jacobi's linear differential equation associated with the second variation.   It is
interesting to see that the initial value procedure involves
the test for conjugate points in such an essential way.

6.   EXTENSIONS AND GENERALIZATIONS

Before entering into a detailed discussion of other
aspects of the theory just presented, let us briefly mention
some possible extensions of the preceding results to more
general quadratic and non-quadratic variational processes.

The minimization of the functional

$$J = \int_0^x [r^2(t) - 2g(t)r(t) -$$

$$\int_0^x k(t, y)r(y) \, dy \, r(t)] \, dt \quad (1)$$

leads to the Euler equation

$$r(t) = g(t) + \int_0^x k(t, y)r(y) \, dy, \qquad 0 \le t \le x. \quad (2)$$

This is a Fredholm integral equation which has been shown to possess a suitable initial value formulation under very general hypotheses on the kernel k.

One way to deal with the problem of a non-quadratic cost functional is to approximate it by a sequence of quadratic problems. If we have the functional

$$I(W) = \int_a^T h(\dot{W}, W, y) \, dy, \, W(a) = c, \quad (3)$$

expanding h and keeping terms only through second-order leads to a problem of the form just treated. Using the Cauchy method, the quadratic variational problem is solved yielding a new approximation. The process is continued by expanding the integrand about the new approximation. In the remainder of this chapter we shall discuss alternate approaches for dealing with the non-quadratic problem directly.

7. "SEMI-QUADRATIC" PROCESSES

Intermediate between the relatively simple quadratic

problem and the full-blown, general variational problem is

a significant class of problems which we shall term "semi-

quadratic". This class is represented by the problem

$$\min_W \; J(W) \; = \; \int_a^T \; [\tfrac{1}{2} \, \dot{W}^2(y) \; + \; F(W, \; y)] \; dy, \quad a < T, \qquad (1)$$

$$W(a) \; = \; c, \; W(T) \; = \; \text{free}. \qquad (2)$$

Many problems of mathematical physics have this form where

F represents a potential energy function and the quadratic

term is the kinetic energy of a dynamical system.

Let us apply our previous procedures to develop an

initial value system characterizing the function u which

minimizes (1) subject to condition (2). Due to the possibly

non-quadratic form of F, we must now take into account the

dependence of u upon c as well as upon a. Consequently,

we shall write

$$u \; = \; u(y, \; a, \; c), \quad a \leq y \leq T, \; |c| < \infty. \qquad (3)$$

Introducing an arbitrary variation function η for which

$\eta(a) = 0$, we write

$$W = u + \varepsilon\eta, \quad \varepsilon \quad \text{a parameter.}$$

Substituting (4) into (1) and arguing in the manner above, we find that

$$\int_a^T [\ddot{u}\eta + F_1(u, y)\eta] \, dy = 0, \tag{5}$$

where

$$F_1(u, y) = \frac{\partial F(u, y)}{\partial u}, \quad (\cdot) = \frac{d}{dy}. \tag{6}$$

Choosing the same variation function η as given in Eq. (3.2) ($\eta = \min(t - a, y - a)$), a nonlinear integral equation for the function u is obtained

$$u(t, a, c) = c - \int_a^T F_1(u, y)\eta \, dy,$$

$$a \le t, \, y \le T, \, |c| < \infty. \tag{7}$$

In Eq. (7) use was made of the condition (2).

The minimizer u is to be studied as a function of c and a. Throughout, the position of the right end, T, is held fixed. Differentiate both sides of Eq. (7) with respect to c and then with respect to a to obtain the equations

$$u_c(t, a, c) = 1 -$$

$$\int_a^T \eta(t, y, a)F_{11}(u, y)u_c(y, a, c) \, dy, \quad (8)$$

$$u_a(t, a, c) = \int_a^T F_1(u, y) \, dy -$$

$$\int_a^T \eta(t, y, a)F_{11}(u, y)u_a(y, a, c) \, dy \quad (9)$$

where the obvious notation

$$F_{11}(u, y) = \frac{\partial^2 F(u, y)}{\partial u^2} \quad (10)$$

has been employed.

Be regarding these as linear functional equations of Fredholm type for the functions u_a and u_c and comparing the inhomogeneous terms, it is seen that

$$u_a(t, a, c) = \left(\int_a^T F_1(u, y) \, dy \right) u_c(t, a, c). \quad (11)$$

Introduce the new function r by means of the definition

$$r(c, a) = -\int_a^T F_1(u(y, c, a)y) \, dy, \quad a \leq T, \quad (12)$$

$$|c| < \infty.$$

Through differentiation, it is seen that

$$r_c(c, a) = -\int_a^T F_{11}(u, y)u_c \, dy \tag{13}$$

and

$$r_a(c, a) = F_1(c, a) - \int_a^T F_{11}(u, y)u_a \, dy. \tag{14}$$

Making use of Eqs. (11) and (12), Eq. (14) becomes

$$r_a(c, a) = F_1(c, a) - rr_c, \quad a < T. \tag{15}$$

From Eq. (12) it is seen that

$$r(c, T) = 0, \tag{16}$$

while the initial condition for Eq. (11) is given by

$$u(t, t, c) = c, \quad t \le T. \tag{17}$$

The initial value problem may now be stated. The function r is determined from the quasilinear partial differential equation

$$r_a = F_1(c, a) - rr_c, \quad a < T, \; |c| < \infty, \tag{18}$$

and the initial condition

$$r(c, T) = 0. \tag{19}$$

The function $u(t, a, c)$ is determined by the partial differential equation

$$u_a(t, a, c) = -r(c, a)u_c(t, a, c),$$

$$a \leq t \leq T, \quad |c| < \infty \quad (20)$$

and the initial condition

$$u(t, t, c) = c. \qquad (21)$$

For the function F_1 sufficiently regular and the interval
$|a - T|$ sufficiently small, this Cauchy problem certainly
has a unique solution.

It will be a simple task for the reader to verify
that, in the case where F is quadratic in u, the equations
given here reduce to those given above for the completely
quadratic case.

The numerical procedure outlined in Section 4 still
carries over word-for-word except that now first order partial
differential equations must be solved rather than ordinary
differential equations. As indicated in Section 6, various
approximation techniques can be used. References are given at
the end of the paper to a number of works where both novel and
standard computational techniques for the solutions of such
equations are discussed.

8.   EULER EQUATIONS AND MINIMALITY

At this juncture it is important for us to take up
the question of whether or not the function u, determined as
the unique solution of the initial value problem just stated,
satisfies the Euler equation associated with the functional
(7.1). This equation is

$$\ddot{u}(t, a, c) = F_1(u(t, c, a), t), \quad a \leq t \leq T, \quad (1)$$

$$u(a, a, c) = c, \quad \dot{u}(T, a, c) = 0 \quad (2)$$

for $|a - T|$ sufficiently small.

We will first show that the left and right sides of
Eq. (1), viewed as functions of a and c, t being a fixed
parameter, satisfy the partial differential equation

$$Z_a = -r(c, a)Z_c, \quad a \leq t \leq T, \quad (3)$$

and the initial condition

$$Z(c, t) = F_1(c, t), \quad |c| < \infty. \quad (4)$$

The assumption of a unique solution to the Cauchy system then
insures equality in Eq. (1).

Differentiation of Eq. (7.20) twice with respect to
the parameter t shows that the function $\ddot{u}(t, a, c)$

satisfies Eq. (3). Consider the initial condition of Eq.
(7.21),

$$u(t, t, c) = c. \tag{5}$$

Differentiate with respect to t to obtain

$$\dot{u}(t, t, c) + u_2(t, t, c) = 0. \tag{6}$$

The dot represents differentiation with respect to the first
argument, the subscript 2 represents differentiation with
respect to the second argument. According to Eq. (7.20), Eq.
(6) becomes

$$\dot{u}(t, t, c) - r(c, t)u_c(t, t, c) = 0. \tag{7}$$

Keeping Eq. (5) in mind, we see that

$$\dot{u}(t, t, c) = r(c, t). \tag{8}$$

Differentiation of Eq. (8) with respect to t then yields

$$\ddot{u}(t, t, c) + \dot{u}_2(t, t, c) = r_t(c, t). \tag{9}$$

Next, differentiate Eq. (7.20) with respect to t to see that

$$\dot{u}_2(t, t, c) = -r(c, t)\dot{u}_c(t, t, c)$$

$$= -r(c, t)r_c(c, t). \tag{10}$$

Since

$$r_t(c, t) = F_1(c, t) - r(c, t)r_c(c, t), \tag{11}$$

Eq. (9) becomes

$$\ddot{u}(t, t, c) = F_1(c, t), \tag{12}$$

which is the desired initial condition of the function $\ddot{u}$ at $a = t$.

Now consider the right side of Eq. (1)

$$W(c, a) = F_1(u(t, c, a), t), \quad a < t, \tag{13}$$

again viewing t as merely a parameter. It is clear that

$$W(c, t) = F_1(u(t, t, c), t)$$

$$= F_1(c, t). \tag{14}$$

Through differentiation with respect to a and c, Eq. (13) becomes

$$W_a = F_1(u, t)u_a$$

$$= F_1(u, t)[-ru_c], \tag{15}$$

$$W_c = F_1(u, t)u_c. \tag{16}$$

It follows that W satisfies the desired partial differential

equation

$$W_a = -rW_c. \tag{17}$$

This completes the proof of the Euler equation.

The first boundary condition in Eq. (2) is given in Eq.

(7.21).

Finally, consider the free boundary condition in Eq.

(2). Differentiate Eq. (7.20) with respect to t and put

t = T to obtain the homogeneous partial differential equa-

tion

$$\dot{u}_a(T,\ a,\ c) = -r(c,\ a)\dot{u}_c(T,\ a,\ c), \qquad a < T. \tag{18}$$

Notice also that

$$\dot{u}(T,\ T,\ c) = r(c,\ T) = 0. \tag{19}$$

Consequently, by uniqueness we have

$$\dot{u}(T,\ a,\ c) = 0. \tag{20}$$

Thus, the solution of the Cauchy system does satisfy the

Euler equation (1) and the associated boundary conditions (2).

9. THE BELLMAN-HAMILTON-JACOBI EQUATION

As an added test of the validity of the Cauchy system

derived in Section 7, we now show that the function u,

produced by solving the initial value system, when substituted

into the functional J of Eq. (7.1) leads to a solution of

the Bellman-Hamilton-Jacobi equation.

Introduce the function f by the relation

$$f(c, a) = \min_{w} J(w). \tag{1}$$

Use of the Principle of Optimality leads directly to the

partial differential equation

$$-f_a = F(c, a) - (1/2)f_c^2, \quad a < T. \tag{2}$$

The initial condition at a = T is

$$f(c, T) = 0. \tag{3}$$

Eq. (2) is the B-H-J equation for the functional (7.1). We

must show that if the function u is used in (7.1), the

resulting functional satisfies Eqs. (2) and (3).

Let the function I be defined by

$$I(c, a) = J(u), \tag{4}$$

where u is determined by Eqs. (7.18)-(7.21). By using Eq.

(7.11) and the Cauchy system of Section 7, it is seen that

$$-I_a = (1/2)r^2(c, a) + F(c, a) + r(c, a)I_c. \tag{5}$$

It remains but to see that

$$-I_c = r(c, a). \tag{6}$$

Introduce the function M to be

$$M(c, a) = -J_c = -\int_a^T [uu_c + F_1 u_c] \, dy. \tag{7}$$

Forming the derivatives of M, we have

$$M_c = -\int_a^T [u\ddot{u}_{cc} + (\dot{u}_c)^2 + F_1 u_{cc} + F_{11}(u_c)^2] \, dy, \tag{8}$$

$$M_a = \dot{u}(a, a, c)\dot{u}_c(a, a, c) + F_1(c, a)u_c(a, a, c) -$$

$$\int_a^T [\ddot{u}u_{ac} + \dot{u}_c\dot{u}_a + F_1 u_{ac} + F_{11}u_a u_c] \, dy. \tag{9}$$

It is now a simple matter to verify that M satisfies the linear partial differential equation

$$M_a = rr_c + F_1(c, a) - rM_c - r_c M, \tag{10}$$

with the initial condition

$$M(c, T) = 0. \tag{11}$$

The unique solution of this system is

$$M(c, a) = r(c, a), \quad a \leq T, \tag{12}$$

which establishes our result.

10.   OPTIMAL CONTROL PROCESSES

With the advent of the digital computer and the space-age, a great deal of effort has been expended in reformulating optimal control problems as initial value problems. By contrast, commonly employed indirect methods of the calculus of variations and the Pontryagin theory require the solution of boundary-value problems. Frequently, these problems are difficult to solve numerically and are often unstable.

In this section, we make use of our earlier ideas to derive an initial value problem whose solution furnishes the extremal arc of the optimal control problem. This development is an extension of our earlier ideas to a variational problem having a differential equation side constraint.

In subsequent sections, we shall attack the question of state and control variable constraints.

The problem we now wish to consider is

$$\min \, I(w) = \int_{a}^{T} f(t, \, y, \, w) \, dt \qquad (1)$$

subject to the differential constraint

$$\frac{dy}{dt} = g(t, \, y, \, w), \quad a \le t \le T, \qquad (2)$$

$$y(a) = c. \qquad (3)$$

Let

$$u = u(t, \, a, \, c), \quad a \le t \le T, \qquad (4)$$

be the minimizing control and

$$x = x(t, \, a, \, c), \quad a \le t \le T \qquad (5)$$

be the minimizing trajectory. As before, we shall assume
the final time T is fixed. We will now derive a Cauchy
system for determining x and u.

11. PRELIMINARY DERIVATIONS

An arbitrary admissible variation in the control of
the form

$$w = u + \varepsilon v \qquad (1)$$

and in the corresponding trajectory

$$y = x + \varepsilon\eta, \tag{2}$$

where ε is an arbitrary parameter, leads to the necessary condition

$$\int_a^T [f_y(t, x, u)\eta + f_w(t, x, u)v] \, dt = 0 \tag{3}$$

with the constraint

$$\dot{\eta} = g_x(t, x, u)\eta + g_u(t, x, u)v, \tag{4}$$

$$\eta(a) = 0. \tag{5}$$

From Eq. (4) it is evident that for an arbitrary function $\lambda(t)$,

$$\lambda(t)[-\dot{\eta} + g_y(t, x, u)\eta + g_w(t, x, u)v] = 0. \tag{6}$$

Adding Eq. (6) to the integrand in Eq. (3) leads to

$$\int_a^T \{[f_y(t', x, u) + g_y(t', x, u)\lambda]\eta - \dot{\eta}\lambda +$$

$$[f_w(t', x, u) + g_w(t', x, u)\lambda]v\} \, dt' = 0. \tag{7}$$

In particular, we can choose λ such that the coefficient of v in Eq. (7) is zero, i.e.,

$$f_w(t', x, w) + g_w(t', x, w)\lambda \Big|_{w=u} = 0,$$

$$a \leq t' \leq T. \quad (8)$$

For λ satisfying relation (8), Eq. (7) yields

$$\int_a^T \{[f_y(t', x, u) +$$

$$g_y(t', x, u)\lambda]\eta - \dot{\eta}\lambda \, dt'\} = 0. \quad (9)$$

Since the variation η may depend upon a and c as well as upon t, differentiating Eq. (9) with respect to a and c gives the relation

$$[\{f_y(t, x, u) + g_y(t, x, u)\lambda\}\eta - \dot{\eta}\lambda]_{t=a} +$$

$$\int_a^T [\{f_{yx}x_a + f_{yu}u_a + (g_{yx}x_a + g_{yu}u_a)\lambda +$$

$$g_y\lambda_a\}\eta + \{f_y + g_y\lambda\}\eta_a -$$

$$(\eta_a\lambda + \dot{\eta}\lambda_a)] \, dt = 0, \quad (10)$$

$$\int_a^T [\{f_{yx}x_c + f_{yu}u_c + (g_{yx}x_c +$$

$$g_{yu}u_c)\lambda + g_y\lambda_c\}\eta + (f_y + g_y\lambda)\eta_c -$$

$$(\dot{\eta}_c\lambda + \dot{\eta}\lambda_c)] \, dt = 0. \quad (11)$$

These relations will be of use in our subsequent development.

12.   TRANSLATIONAL INVARIANCE

In order to determine the relevant equations satisfied by the optimizing functions   x   and   u, we now discuss
a fundamental imbedding principle which we shall call the
Principle of Translational Invariance.

Consider the optimal trajectory which begins at some
point   (a, c)   and continues to a terminal point at time   T.
In moving from the point on the trajectory at time   a   to
the point corresponding to time $a + \Delta$, we have the following changes taking place

$$a \rightarrow a + \Delta, \tag{1}$$

$$c \rightarrow c + \dot{x}(a, a, c)\Delta, \tag{2}$$

to terms $0(\Delta^2)$. This observation allows us to write

$$u(t, a, c) = u(t, a + \Delta, c +$$
$$\dot{x}(a, a, c)\Delta) + 0(\Delta^2), \tag{3}$$

$$x(t, a, c) = x(t, a + \Delta, c +$$
$$\dot{x}(a, a, c)\Delta) + 0(\Delta^2) \tag{4}$$

where

$$\Delta > 0, \ a + \Delta \le t < T.$$

Eqs. (3) and (4) express a simple translational invariance

property possessed by the optimal control and optimal trajec-

tory. For ease of writing, let us denote the initial slope of

the optimal trajectory by p,

$$p(a, c) = \dot{x}(a, a, c). \tag{5}$$

Then the limiting forms as $\Delta \to 0$ of Eqs. (3) and (4),

assuming the requisite smoothness properties, are

$$u_a(t, a, c) = -p(a, c)u_c(t, a, c), \tag{6}$$

$$x_a(t, a, c) = -p(a, c)x_c(t, a, c), \quad a \le t \le T. \tag{7}$$

Application of the same principle to the function λ yields

$$\lambda_a(t, a, c) = -p(a, c)\lambda_c(t, a, c), \quad a \le t \le T. \tag{8}$$

The initial condition for x at a = t is

$$x(t, t, c) = c. \tag{9}$$

The initial conditions on u and λ will be specified later.

13. EQUATIONS FOR p

From the governing differential equation (10.2), we

know that

$$\dot{x}(t, a, c) = g(t, x(t, a, c), u(t, a, c)). \qquad (1)$$

Consequently, letting $t = a$, we have

$$p(a, c) = g(a, c, q(a, c)) \qquad (2)$$

where we have defined

$$q(a, c) = u(a, a, c). \qquad (3)$$

To obtain an equation for q, recall Eq. (11.8) with $t = a$:

$$f_w(a, c, q(a, c)) +$$

$$g_w(a, c, q(a, c))\lambda(a, a, c) = 0. \qquad (4)$$

Thus, if an expression for $\lambda(a, a, c)$ can be obtained our system will be complete since q will then be obtainable from Eq. (4), while p will then be obtained from Eq. (2).

For convenience, define

$$\lambda(a, a, c) = \gamma(a, c). \qquad (5)$$

From substituting the equations (12.6) and (12.7) for x and u into Eq. (11.10), and then substituting from Eq. (11.11) into the resulting expression and simplifying terms by choosing the variation η as

$$\eta = t - a, \quad a \le t \le T, \tag{6}$$

the resulting expression is

$$\lambda(a, a, c) - \int_a^T [p(a, c)\lambda_c -$$

$$(f_y + g_y\lambda) + \lambda_a \, dt] = 0. \tag{7}$$

Equation (12.8) allows us to reduce this to

$$\gamma(a, c) = \lambda(a, a, c) = \int_a^T (f_y + g_y\lambda) \, dt. \tag{8}$$

Next, differentiate both sides of Eq. (7) with respect to a and c to obtain the relations

$$\gamma_a(a, c) = -[f_y(a, x(a, a, c), u(a, a, c)) +$$

$$g_y(a, x(a, a, c), u(a, a, c)) \times$$

$$\lambda(a, a, c)] + \int_a^T [f_{yx}x_a + f_{yu}u_a +$$

$$(g_{yx}x_a + g_{yu}u_a)\lambda + g_y\lambda_a] \, dt, \tag{9}$$

$$\gamma_c(a, c) = \int_a^T [f_{yx}x_c + f_{yu}u_c +$$

$$(g_{yx}x_c + g_{yu}u_c)\lambda + g_y\lambda_c] \, dt. \tag{10}$$

Using the definitions of p and q, as well as the initial

condition on x, Eq. (9) becomes

$$\gamma_a(a, c) = -[f_y(a, c, q(a, c)) +$$

$$g_y(a, c, q(a, c))\gamma(a, c)] -$$

$$p(a, c)\int_a^T [f_{yx}x_c + f_{yu}u_c +$$

$$(g_{yx}x_c + g_{yu}u_c)\lambda + g_y\lambda_c]\ dt. \quad (11)$$

Noting Eq. (10), we see that this is

$$\gamma_a(a, c) + p(a, c)\gamma_c(a, c) +$$

$$g_y(a, c, q)\gamma(a, c) + f_y(a, c, q) = 0. \quad (12)$$

From Eq. (8) it is evident that

$$\gamma(T, c) = 0. \qquad\qquad\qquad (13)$$

14.   RECAP OF THE CAUCHY SYSTEM

     Since the foregoing derivation was rather involved,

let us now summarize the Cauchy system for determining the

optimal trajectory, x, and the optimal control u. The equa-

tions for x and u are

$$x_a(t, a, c) = -p(a, c)x_c(t, a, c), \qquad\qquad (1)$$

$$u_a(t, a, c) = -p(a, c)u_c(t, a, c), \quad a \leq t \leq T. \quad (2)$$

The initial conditions at $a = t$ are

$$x(t, t, c) = c, \quad (3)$$

$$u(t, t, c) = q(t, c). \quad (4)$$

The auxiliary functions p and q are determined by the following partial differential equation and finite relations

$$\gamma_a(a, c) = -p(a, c)\gamma_c(a, c) +$$

$$g_y(a, c, q)\gamma(a, c) + f_y(a, c, q), \quad a \leq t \leq T, \quad (5)$$

$$\gamma(T, c) = 0, \quad (6)$$

$$f_w(a, c, q) + g_w(a, c, q)\gamma(a, c) = 0, \quad (7)$$

$$p(a, c) = g(a, c, q). \quad (8)$$

To avoid confusion, it must be remembered that g_y denotes the partial of g with respect to its second argument and g_w refers to differentiation with respect to the third argument. Similarly, for f_y and f_w.

It is a straightforward matter to show that if the governing differential equation is linear and the integrand in Eq. (10.1) is quadratic, the initial value problem just given reduces to a system of ordinary differential equations.

Also, the extension to the vector-matrix case can be made in a simple manner. However, care must be taken in obtaining the functions p and q since, in general, sets of nonlinear equations need to be solved.

15. QUADRATIC PROCESSES & LINEAR EQUATIONS

To illustrate the above ideas, consider minimizing the quadratic cost function

$$I(w, y) = \int_a^T (\frac{1}{2} w^2 + g(t)y^2 \, dt], \tag{1}$$

subject to the linear dynamics

$$\frac{dy}{dt} = b(t)y + d(t)w, \quad y(a) = c. \tag{2}$$

The functions f and g of the previous section now take the form

$$f(t, y, w) = \frac{1}{2} (w^2 + g(t)y^2), \tag{3}$$

$$g(t, y, w) = b(t)y + d(t)w. \tag{4}$$

Equation (14.8) for $p(a, c)$ becomes

$$p = b(a)c + d(a)q(a, c) \tag{5}$$

while Eq. (14.7) for q is

$$q(a, c) + d(a)\gamma(a, c) = 0. \tag{6}$$

The linear nature of the problem now allows us to separate variables in the partial differential equations for x, u, and γ. We then write

$$x(t, a, c) = cX(t, a), \tag{7}$$

$$u(t, a, c) = cU(t, a), \tag{8}$$

$$\gamma(a, c) = cG(a). \tag{9}$$

The equations satisfied by X, U, and G are

$$X_a(t, a) = [d^2(a)G(a) - b(a)]X(t, a), \tag{10}$$

$$U_a(t, a) = [d^2(a)G(a) - b(a)]U(t, a), \tag{11}$$

$$G'(a) = -g(a) + d^2(a)G^2(a). \tag{12}$$

The initial conditions are

$$X(t, t) = 1, \tag{13}$$

$$U(t, t) = -d(t)G(t), \tag{14}$$

$$G(T) = 0. \tag{15}$$

Note that Eqs. (10)-(15) comprise a set of ordinary differential equations with prescribed initial conditions, t being merely a parameter.

16.   CONSTRAINED CONTROL AND THE PONTRYAGIN THEORY

In the remaining sections, we shall add one more degree of complexity to the processes we have been treating by introducing a control process with constraints on the control variable. This is a very natural constraint in many instances, often corresponding to the case when the control variable represents an amount of resource expended, the constraint then amounting to the statement that we cannot spend more than we possess to control the system. Clearly, there are many types of constraints of this nature that one could impose. However, we shall be content with the case in which a uniform bound on the magnitude of the control is specified. After dealing with this case, we shall then indicate promising directions to follow in dealing with other situations such as time-dependent constraints, state variable constraints, and so forth.

The class of problems we treat in detail is a type of constrained optimal control problem normally treated by the

Pontryagin Maximum Principle. A Cauchy problem is obtained

which, in some ways, may be viewed as an extension of the

classical Hamilton-Jacobi theory to constrained variational

processes.

Let the state of a system at time   t   be denoted by

y, a scalar, and let its dynamical equations of motion by

$$\dot{y} = g(y, w), \quad a \leq t \leq T, \tag{1}$$

$$y(a) = c, \quad |c| < \infty, \tag{2}$$

where   w   is a control variable. We assume that the control

is subject to the constraint

$$|w| \leq M, \quad a \leq t \leq T, \quad M \text{ a real scalar.} \tag{3}$$

The cost functional to be minimized is

$$I = \int_a^T f(y, w) \, dt. \tag{4}$$

The objective, of course, is to select a control function   w,

subject to the constraint (3), which minimizes   I. We shall

assume that the functions   f   and   g   are sufficiently smooth

and the interval   [a, T]   sufficiently small so that the

indicated operations in our subsequent discussion are valid.

According to the Pontryagin theory, the solution to

the above problem is obtained by introducing the variable λ and function H such that

$$H(y, w, \lambda) = f(y, w) + \lambda g(y, w) \tag{5}$$

and solving the two point boundary value problem

$$\dot{y} = H_\lambda = g(y, w), \tag{6}$$

$$-\dot{\lambda} = H_y = f_y(y, w) + \lambda g_y(y, w), \qquad a \le t \le T, \tag{7}$$

$$y(a) = c, \tag{8}$$

$$\lambda(T) = 0, \tag{9}$$

where

$$\min_{|z| \le M} H(y, z, \lambda) = H(y, w, \lambda). \tag{10}$$

Our aim is to produce a solution of these equations in the form of an appropriate Cauchy system.

17.   DERIVATION OF THE CAUCHY SYSTEM

Hold the terminal time T fixed, and consider the initial epoch a, $a \le T$, and the initial state c, $|c| < \infty$, to be variable. Introduce the minimal cost function F,

$$F(c, a) = \min_{|w| \leq M} \int_a^T f(y, w) \, dt, \tag{1}$$

where the state and control variables are subject to the dynamical equation (16.1) and the initial condition (16.2). According to Bellman's Principle of Optimality, the function F satisfies the partial differential equation

$$F_a(c, a) = \min_{|z| \leq M} [f(c, z) + g(c, z)F_c(c, a)],$$

$$a \leq T, \tag{2}$$

and the initial condition

$$F(c, T) = 0. \tag{3}$$

Furthermore, for each value of c and a, let us denote the minimizing value of z by

$$z_{\min} = r(c, a), \qquad a \leq T, \ |c| < \infty. \tag{4}$$

In order to deal with the functions y, w, and λ as functions of c and a, let us again write

$$y = y(t, a, c) \tag{5}$$

$$w = w(t, a, c) \tag{6}$$

$$\lambda = \lambda(t, a, c) \tag{7}$$

Then application of the Principle of Translation Invariance, enunciated above, gives

$$y(t, a, c) = y(t, a + \Delta, c + g(c, r(c, a))\Delta) + 0(\Delta). \tag{8}$$

The limiting form of (8) is

$$y_a(t, a, c) = -g(c, r(a, c))y_c(t, a, c),$$

$$a < t < T, \ |c| < \infty. \tag{9}$$

The functions w and λ also satisfy the same partial differential equation,

$$w_a(t, a, c) = -g(c, r(c, a))w_c(t, a, c), \tag{10}$$

$$\lambda_a(t, a, c) = -g(c, r(c, a))\lambda_c(t, a, c). \tag{11}$$

The initial conditions on y, w, and λ are

$$\dot{y}(t, t, c) = c, \tag{12}$$

$$w(t, t, c) = r(t, c), \tag{13}$$

$$\lambda(t, t, c) = F_c(c, t), \ |c| < \infty. \tag{14}$$

18.   REMARKS

It is clear from the previous sections that in order to deal with the constrained control process, we have had to lower our sights somewhat and settle for a formulation which makes use of dynamic programming to obtain a basic auxiliary function. Hopefully, future work will show that this function can be obtained independently and without recourse to other optimization methods. However, the formulation given does serve to underscore the point that a combination of methods can be expected to give better results than rigid adherence to one orthodoxy or another. With dynamic programming, the function r is easy while w is hard; conversely, for invariant imbedding we can easily get the trajectory w but an equation for r seems difficult. Hence, combination of the two methods gives a seemingly practical algorithm. Presumably, the same idea can be applied to problems involving state variable inequality constraints, random effects, etc.

CHAPTER FIVE

NOTES AND REFERENCES

§1.   For a discussion of standard optimization methods

       such as the calculus of variations, dynamic program-

       ming, and the maximum principle see

           Gelfand, I. M., and S. V. Fomin, Calculus of

           Variations, Prentice-Hall, Inc., Englewood Cliffs,

           N. J., 1963,

           Bellman, R. and S. Dreyfus, Applied Dynamic

           Programming, Princeton University Press, Princeton,

           N. J., 1962,

           Pontryagin, L. S., et al., The Mathematical Theory

           of Optimal Processes, John Wiley & Sons, New York,

           1962.

§2.   An in-depth treatment of the quadratic criterion-

       linear equation control problem may be found in

           Bellman, R., Introduction to the Mathematical

           Theory of Control Processes, Vol. 1, Academic

           Press, New York, 1967.

§6.   The treatment of a non-quadratic cost functional by

       approximation is discussed in

Bellman, R., and R. Kalaba, <u>Quasilinearization and</u>

<u>Nonlinear Boundary Value Problems</u>, American

Elsevier Co., New York, 1965.

§7-9.   These results were first given in Casti, J., R.

Kalaba, and B. Vereeke, "Invariant Imbedding and a

Class of Variational Problems,: <u>J. Opt. Theory &</u>

<u>Appl.</u>, Vol. 3, No. 2, Feb. 1969.

An interesting treatment of the use of variational

theory in mechanics is given by

Lanczos, C., <u>The Variational Principles of</u>

<u>Mechanics</u>, 3rd ed., University of Toronto Press,

Toronto, 1966.

§12.   The Principle of Translational Invariance may be

thought of as a manifestation of the fundamental

semi-group property of the unique solution to an

initial value problem.

§15.   The definitive works on the application of dynamic

programming to control processes are

Bellman, R., <u>Adaptive Control Processes</u>, Princeton

University Press, Princeton, N. J., 1961,

Dreyfus, S., <u>Dynamic Programming and the Calculus</u>

<u>of Variations</u>, Academic Press, New York, 1965.

Computational aspects of dynamic programming together

with numerous examples from many areas are in

    Larson, R., <u>State Increment Dynamic Programming</u>,

American Elsevier Co., New York, 1970.

CHAPTER SIX

APPLICATIONS IN THE PHYSICAL SCIENCES

1. INTRODUCTION

The test of any theory is how applicable it is to the solution of broad classes of significant problems. In this chapter we wish to shift our emphasis from the development of theories to their application in concrete physical situations. As is usually the case, the physical problems we deal with do not always match the foregoing theory in a one-to-one correspondence. As a result, we have occasion to introduce minor extensions and variations to the preceding results leaving the fundamental ideas unchanged. These variations, however, are useful in their own right since they serve to amplify and extend the previous theory as well as illustrate useful techniques in dealing with particular problems that arise.

229

The examples given below were selected for their diversity as well as their intrinsic physical interest and are meant to be represenaative of the classes and types of problems which can be handled. We wish to show the generality of the imbedding concept in many different situations and dispel the notion that it is a technique which is of limited utility in a specialized area such as neutron transport theory.

The previous sentence not withstanding, our first example is from radiative transport theory in the atmosphere. We consider the problem of determining the source function for a plane-parallel atmosphere bounded by an absorber and also the case when it is bounded by a specular reflector. We show that the ideas of Chapter 4 give a very satisfactory resolution to both these problems.

The second application is from the field of analytical mechanics. In the light of Chapter 3, we re-examine Hamilton's equations of motion for a dynamical system and deduce some new equations of mechanics. These equations are then applied to the classical harmonic oscillator as an illustrative example of their use. Then thin beams are studied.

Filtering a known signal in the presence of noise is a problem of considerable interest in many areas of scientific activity. We show how Fredholm integral equations play

a central role in many such investigations and how the
theory developed in Chapter 4 can be used to obtain new on-
line filtering schemes. We also indicate applications of
our ideas to the detection of signals in the presence of
colored noise.

In the study of electromagnetic fields in metallic
slabs and semiconductors of finite thickness, account must
be taken of non-local wave interactions. Our last example
provides a Cauchy system suitable for treating the integro-
differential equations of non-local wave interaction.

2. THE SOURCE FUNCTION

In the study of planetary radiation fields a central
position is occupied by the source function. Knowledge of
the source function is sufficient to determine reflected,
transmitted, and internal intensities of radiation as well
as other quantities of physical interest. For these reasons,
we will show how the imbedding theory derived earlier can be
used to obtain Cauchy systems suitable for obtaining this
basic function.

Consider an homogeneous, plane-parallel atmosphere of
finite thickness x, illuminated at the top by parallel rays
of net flux π, the rays making an angle arccos z with

respect to the downward normal, $0 \le z \le 1$. Assume the medium

absorbs radiation and scatters it isotropically, the albedo

for single scattering being λ. Denote the source function

at the altitude t by $J(t, x, z)$, $0 \le t \le x$. Physically,

J is the rate of production of particles per unit volume

per unit solid angle at the altitude t for a medium of

thickness x due to illumination by parallel rays in the

direction arccos z.

Assuming a Poisson distribution for the probability

of a particle interacting in a small slab of thickness

$\Delta \ll 1$, a simple argument shows that J satisfies the

integral equation

$$J(t, x, z) = (\lambda/4)e^{-(x-t)/z} +$$

$$(\lambda/2)\int_0^x E_1(|t - y|)J(y, x, z) \, dy,$$

$$0 \le t \le x, \ 0 \le \lambda, \ z \le 1. \quad (1)$$

Equation (1) assumes the atmosphere is bounded by a perfect

absorber. Physically, the first term on the right hand

side of (1) comes about from incident radiation which goes

directly to the altitude t without prior interaction,

interacting for the first time at t. The integral term

occurs when particles are produced at the altitude y, go to

the altitude t without interaction, and then interact at t.

The function $E_1(r)$ is the probability of traversing a distance $r > 0$ without interaction, and is a function only of the distance between the two points involved.

3. INITIAL VALUE SYSTEM FOR J

Since the kernel in Eq. (2.1) may be expressed as

$$\frac{\lambda}{2} E_1(|t - y|) = \frac{\lambda}{2} \int_0^1 e^{-|t-y|/z} \, dz/z, \tag{1}$$

we see that the integral equation satisfied by J falls into the class of equations treated in Chapter 4. Recalling Eq. (2.2) of Chapter 4, we see that the appropriate weighting function for the Cauchy system given there is

$$w(z) = \frac{\lambda}{2z} . \tag{2}$$

Due to the linearity of Eq. (2.1), the source function J of the current discussion is obtained from the auxiliary function J of Chapter 4 by a simple factor of $(\lambda/4)$.

Combining Eq. (2) with the previous remark, we see that the appropriate Cauchy system for the source function is

$$X_x(x, z) = (\lambda/2)Y(x, z)\int_0^1 Y(x, z') \, dz'/z', \tag{3}$$

$$Y_x(x, z) = -\frac{1}{z} Y(x, z) +$$

$$(\lambda/2)X(x, z)\int_0^1 Y(x, z') \, dz'/z', \quad (4)$$

$$J_x(t, x, z) = -\frac{1}{z} J(t, x, z) +$$

$$(\lambda/2)X(x, z)\int_0^1 J(t, x, z') \, dz'/z', \quad (5)$$

$$X(0, z) = 1, \quad (6)$$

$$Y(0, z) = 1, \quad (7)$$

$$J(t, t, z) = (\lambda/4)X(t, z), \quad 0 \le t \le x,$$

$$0 \le \lambda, z \le 1. \quad (8)$$

Equations (3)-(8) may be reduced to a set of ordinary differential equations and integrated numerically by the procedures discussed in Chapter 4.

What is rather remarkable about this formulation is
that the auxiliary functions X and Y, which played a
necessary mathematical role in our earlier discussion, turn
out to have physical meaning in their own right, representing
the rate of production of particles at the top and bottom,
respectively, of the medium. As a result, if only behavior
at the boundary is of concern, no information is necessary
concerning internal activity. Thus, our theory may be

thought of as dealing with what the physicists call

"observables".

4. EXTERNAL AND INTERNAL RADIATION FIELDS

With a means at our disposal to calculate the source

function J we may, as noted above, now obtain various

other functions of interest. For example, let r(v, z, x)

be the intensity of radiation reflected from the top of the

atmosphere of thickness x in the direction arccos v due

to incident radiation in the direction arccos z, and let

t(v, z, x) represent the transmitted intensity. Simple

physical arguments give

$$r(v, z, x) = \frac{1}{v} \int_0^x J(y, x, z)e^{-(x-y)/v} \, dy, \qquad (1)$$

$$t(v, z, x) = \frac{1}{v} \int_0^x J(y, x, z)e^{-y/v} \, dy. \qquad (2)$$

Initial value systems for the functions r and t or,

more properly, for the functions R = vr, T = vt can be

readily obtained. We will discuss these in more detail in a

later section.

Relations similar to (1) and (2) can also be obtained

for the internal intensity functions. Let I(t, $\pm$v, x, z) denote the intensity of the diffuse radiation at the altitude

t in a direction whose direction cosine with respect to the

upward normal is v, x and t having the same meaning as

before. The $\pm$ refers to whether the intensity is in the

upward (+) or downward (-) direction. The relevant equa-

tions are

$$I(t, +v, x, z) = \frac{1}{v} \int_0^t e^{-(t-y)/v} J(y, x, z) \, dy, \qquad (3)$$

$$I(t, -v, x, z) = \frac{1}{v} \int_t^x e^{-(y-t)/v} J(y, x, z) \, dy. \qquad (4)$$

The references at the end of the chapter give Cauchy systems

and numerical results for these quantities.

5.   REFLECTING SURFACES

        Let us now generalize the preceding example to the

case when our atmosphere is bounded at the bottom by a

specular reflector. This case will illustrate the imbedding

approach to more complex boundary conditions. Extensions to

laws of scattering other than isotropic are covered in the

papers referenced at the end of the chapter.

        Consider the same atmosphere as that given above

which is bounded by a specular reflector characterized by the function $\rho(v)$, $0 \le v \le 1$, which represents the probability

that a particle impinging on the bottom surface with angle of

incidence arccos v is specularly reflected. Denoting the

source function by $J(t, x, z)$ as before, straightforward

physical reasoning gives the equation

$$J(t, x, z) = \frac{\lambda}{4} [e^{-(x-t)/z} +$$

$$\rho(z)e^{-(x+t)/z}] + \frac{\lambda}{2} \int_0^x [E_1(|t - y|) +$$

$$f(t + y)]J(y, x, z) \, dy. \quad (1)$$

where

$$E_1(s) = \int_0^1 e^{-s/z'} \, dz'/z',$$

$$f(s) = \int_0^1 e^{-s/z'} \rho(z') \, dz'/z', \quad s > 0.$$

Rather than repeat the prior derivations for J, let us now

show how a suitable Cauchy problem may be obtained for the

reflected and transmitted intensities.

Let

$r(v, z, x)$ = the intensity of the diffusely

reflected radiation with angle of

reflection arccos v.

The reflection function   r   is expressed in terms of the source function   J   by the relation

$$r(v, z, x) = \frac{1}{v} \int_0^x J(y, x, z) [e^{-(x-y)/v} +$$

$$\rho(v) e^{-(x+y)/v}] \, dy. \quad (2)$$

It is convenient to work with the quantity   R   defined as

$$R(v, z, x) = vr(v, z, x) \quad (3)$$

The transmission function τ is defined as

$\tau(v, z, x)$ = the intensity of the diffuse radiation
at the bottom of the atmosphere in a
direction making an angle   arccos v
with respect to a downward normal to
the specular reflector.

In terms of   J, the function τ is expressed as

$$\tau(v, z, x) = \frac{1}{v} \int_0^x e^{-y/v} J(y, x, z) \, dy. \quad (4)$$

We also introduce the function   T   as

$$T(v, z, x) = v\tau(v, z, x) \quad (5)$$

We shall now derive initial value problems for the functions R   and   T, x   being viewed as the independent variable.

6.   REFLECTION AND TRANSMISSION

Differentiating Eq. (5.3) with respect to   x   gives

$$R_x(v, z, x) = J(x, x, z)[1 + \rho(v)e^{-2x/v}] -$$

$$\frac{1}{v} R(v, z, x) + \int_0^x J_x(y, x, z)[e^{-(x-y)/v} +$$

$$\rho(v)e^{-(x+y)/v}] \, dy. \quad (1)$$

To obtain an expression for J_x,   differentiate Eq. (5.1) with
respect to   x   to obtain

$$J_x(t, x, z) = \frac{-\lambda}{4z} [e^{-(x-t)/z} +$$

$$\rho(z)e^{-(x+t)/z}] + \frac{\lambda}{2} [E_1(x - t) + f(x + t)] \cdot$$

$$J(x, x, z) + \frac{\lambda}{2} \int_0^x [E_1(|t - y|) +$$

$$f(t + y)]J_x(y, x, z) \, dy. \quad (2)$$

Comparing Eqs. (3) and (5.1) and making use of the definitions
of E_1 and   f, we have

$$J_x(t, x, z) = -\frac{1}{z} J(t, x, z) +$$

$$2J(x, x, z)\int_0^1 J(t, x, z') \, dz'/z'. \quad (3)$$

Equation (1) becomes

$$R_x(v, z, x) = J(x, x, z)[1 + \rho(v)e^{-2x/v}] -$$

$$\frac{1}{v} R(v, z, x) + \int_0^x [-\frac{1}{z} J(y, x, z) +$$

$$J(x, x, z)\Phi(y, x)][e^{-(x-y)/v} +$$

$$\rho(v)e^{-(x+y)/v}] \, dy, \quad (4)$$

where

$$\Phi(t, x) = 2\int_0^1 J(t, x, z') \, dz'/z'. \tag{5}$$

This equation reduces to

$$R_x(v, z, x) = -(\frac{1}{v} + \frac{1}{z})R(v, z, x) +$$

$$J(x, x, z) \cdot \{1 + \rho(v)e^{-2x/v} +$$

$$\int_0^x \Phi(y, x)[e^{-(x-y)/v} + \rho(v)e^{-(x+y)/v}] \, dy\} \tag{6}$$

Making use of Eqs. (5) and (5.3), it is seen that

$$R_x(v, z, x) = -(\frac{1}{v} + \frac{1}{z})R(v, z, x) +$$

$$J(x, x, z) \cdot [1 + \rho(v)e^{-2x/v} +$$

$$2 \int_0^1 R(v, z', x) \, dz'/z']. \quad (7)$$

To express $J(x, x, z)$ in terms of R, we return to Eq. (5.1) and let $t = x$ obtaining

$$J(x, x, z) = \frac{\lambda}{4} [1 + \rho(z) e^{-2x/z}] +$$

$$\frac{\lambda}{2} \int_0^x [E_1(x - y) + f(x + y)] J(y, x, z) \, dy. \quad (8)$$

Keeping in mind Eq. (5.3), this becomes

$$J(x, x, z) = \frac{\lambda}{4} [1 + \rho(z) e^{-2x/z}] +$$

$$\frac{\lambda}{2} \int_0^1 R(v', z, x) \, dv'/v' = \frac{\lambda}{4} [1 + \rho(z) e^{-2x/z} +$$

$$2 \int_0^1 R(v', u, x) \, dv'/v']. \quad (9)$$

The final differential equation for R is

$$R_x(v, z, x) = -(\frac{1}{v} + \frac{1}{z}) R(v, z, x) +$$

$$\frac{\lambda}{4} [1 + \rho(z) e^{-2x/z} + 2 \int_0^1 R(v', z, x) \, dv'/v'] \cdot$$

$$[1 + \rho(v) e^{-2x/v} + 2 \int_0^1 R(v, z', x) \, dz'/z']. \quad (10)$$

The initial condition at $x = 0$ according to Eq. (5.2) is

$$R(v, z, 0) = 0, \qquad 0 \le v, z \le 1. \tag{11}$$

Knowing $J(x, x, z)$, the equation for $T(v, z, x)$ is obtained very simply. Differentiate Eq. (5.5) with respect to x to obtain

$$T_x(v, z, x) = e^{-x/v} J(x, x, z) +$$

$$\int_0^x e^{-y/v} J_x(y, x, z) \, dy. \tag{12}$$

According to Eq. (3) this becomes

$$T_x(v, z, x) = e^{-x/v} J(x, x, z) +$$

$$\int_0^x e^{-y/v} [- \frac{1}{z} J(y, x, z) +$$

$$J(x, x, z) \Phi(y, x)] \, dy, \tag{13}$$

or

$$T_x(v, z, x) = e^{-x/v} J(x, x, z) -$$

$$\frac{1}{z} T(v, z, x) + J(x, x, z) \int_0^x e^{-y/v} \Phi(y, x) \, dy. \tag{14}$$

Using the definitions of Φ and R given above, it is seen that

$$T_x(v, z, x) = -\frac{1}{z} T(v, z, x) + \frac{\lambda}{4} [1 + \rho(z)e^{-2x/z} +$$

$$2\int_0^1 R(v', z, x) \, dv'/v'] \cdot [e^{-x/v} +$$

$$2\int_0^1 T(v, z', x) \, dz'/z']. \quad (15)$$

Equation (5.4) gives the initial condition at $x = 0$ as

$$T(v, z, 0) = 0, \qquad 0 \le v, z \le 1. \tag{16}$$

7.   SOME NUMERICAL RESULTS

To test the above formulation, calculations were carried out for various thickness x and values of λ . The conservative situation $\lambda = 1.0$, $\rho(x) = 1.0$, is especially important. For this case we may write the conservation law

$$\pi z = \pi z e^{-2x/z} + 2\pi \int_0^1 R(v, z, x) \, dv \tag{1}$$

or

$$\frac{z}{2} [1 - e^{-2x/z}] = \int_0^1 R(v, z, x) \, dv, \tag{2}$$

indicating that the flux input at angle z equals the flux output at angle   z   since there is no absorption loss and

the bottom surface is a perfect mirror reflector.   The

accuracy with which (2) is satisfied gives a good indication

how accurate our entire calculation will be.

Using an Adams-Moulton integration scheme of step

size   0.005   and Gaussian quadrature of order   N = 7   to

approximate all integrals, the results of Table 1 were

obtained when   x = 4.0.

Table 1:

Conservation Law Calculations, x = 4.0
$$(\times \ 10^{-1})$$

Input Angle	Right Side of (2)	Left side of (2)
1	0.1272300	0.1272302
2	0.6461707	0.6461721
3	1.485383	1.485387
4	2.499992	2.500000
5	3.514560	3.514573
6	4.353365	4.353382
7	4.871423	4.871443

Note that the numerical value of the various angles are

determined by computing   arccos z_i, where z_i = the ith

root of $P_7^*(z)$, the shifted Legendre polynomial of order   7.

8.   ANALYTICAL MECHANICS AND HAMILTON'S EQUATIONS

A fruitful source of two point boundary value problems is the study of dynamical mechanical systems. The conservation laws of physics, when applied to such systems, immediately lead to two-point boundary value problems of the type discussed in Chapter 3. In the next few sections we shall point out the interpretation of our earlier results in the context of mechanics, and illustrate our ideas with the simple harmonic oscillator.

Consider the motion of a particle on a line where we characterize the process by a Hamiltonian $H = H(p, q)$. The equations of motion are

$$\dot{q} \;=\; \frac{\partial H}{\partial p} \,, \; q(0) = 0 \tag{1}$$

$$-\dot{p} \;=\; \frac{\partial H}{\partial q} \,, \; p(T) = c, \quad 0 \le t \le T. \tag{2}$$

We desire to obtain the unknown displacement $q(T)$ at time T. Introduce the function

r(c, T) = the displacement at time T, the initial
displacement being zero and the momentum
at time T being c.

In Chapter 3 we showed that if we consider the system

$$\frac{du}{dz} = F(u, v), \ u(0) = 0, \tag{3}$$

$$-\frac{dv}{dz} = G(u, v), \ v(x) = c, \qquad 0 \le z \le x, \tag{4}$$

then the function $r(c, x) = u(z)\big|_{z=x}$, satisfies the partial differential equation

$$\frac{\partial r}{\partial x} = F(r, \ c) + G(r, \ c) \frac{\partial r}{\partial c} \tag{5}$$

amd the initial condition

$$r(c, \ 0) = 0. \tag{6}$$

Applying this result to the system (1) and (2), we find that $r(c, T)$ satisfies

$$\frac{\partial r}{\partial T} = \frac{\partial H(r,c)}{\partial p} + \frac{\partial H(r,c)}{\partial q} \frac{\partial r}{\partial c}, \tag{7}$$

or

$$\frac{\partial r}{\partial T} = \frac{\partial H(r,c)}{\partial c} \tag{8}$$

A simple example of the foregoing is the harmonic oscillator. In this case

$$H(p, \ q) = \frac{1}{2} [p^2/m + kq^2]$$

with   (9)

$$q(0) = 0, \ p(T) = c, \ m \ \text{ and } \ k \ \text{ constants.}$$

For the unknown displacement at time T, Eq. (8) becomes

$$\frac{\partial r}{\partial T} = \frac{\partial H}{\partial c} = \frac{c}{m} + kr \frac{\partial r}{\partial c} . \tag{10}$$

The solution of this equation, subject to (6) is

$$r(c, T) = \frac{c}{\sqrt{km}} \tan\left(\sqrt{\frac{k}{m}} T\right). \tag{11}$$

This provides the desired displacement at time T and also shows that the function r may become infinite for a finite value of T.

9.   NONVARIATIONAL PRINCIPLES IN DYNAMICS

Let us now extend the previous results in much the same manner that Jacobi extended Hamilton's integration theory. The important point of the following discussion is that the approach presented here applies in the general case in which there may be no variational principle underlying Eqs. (8.1)-(8.2).

For simplicity, consider the system of dynamical equations

$$\frac{dq}{dt} = F(q, p), \quad q(0) = w, \tag{1}$$

$$-\frac{dp}{dt} = G(q, p), \quad p(T) = c, \qquad 0 \le t \le T. \tag{2}$$

In our earlier discussion, we showed that the functions

$$r(c, T, w) = q(T) \tag{3}$$

and

$$\tau(c, T, w) = p(0) \tag{4}$$

satisfy the first-order partial differential equations

$$r_T = F(r, c) + G(r, c)r_c, \tag{5}$$

$$\tau_T = G(r, c)\tau_c. \tag{6}$$

In addition, they satisfy the initial conditions

$$r(c, 0, w) = w, \tag{7}$$

$$\tau(c, 0, w) = c. \tag{8}$$

Let us now follow the path of Jacobi to use the functions r and τ to define the solution functions p and q. Let

$$r = r(c, T, \alpha) \tag{9}$$

be a solution of Eq. (5) for arbitrary values of the parameter α, and let $\tau(c, T, \alpha)$ be a solution of Eq. (6), which now involves α by way of r. Then we assert the following: The equations

$$\tau(p, t, \alpha) = \beta, \tag{10}$$

$$r(p, t, \alpha) = u \tag{11}$$

form a system which implicitly defines q and p as functions of t, α, and β with q and p being solutions of Eqs. (1) and (2).

Let us verify this assertion. Upon differentiation with respect to t, Eq. (10) gives

$$\tau_c(p, t, \alpha)\dot{p} + \tau_T(p, t, \alpha) = 0. \tag{12}$$

Substituting from Eq. (6) then yields

$$\dot{p} = -G(q, p) \tag{13}$$

provided $\tau_c \neq 0$. Eq. (13) is one of the desired relations. Upon differentiating Eq. (11) with respect to t, we find

$$r_c(p, t, \alpha)\dot{p} + r_T(p, t, \alpha) = \dot{q} \tag{14}$$

or

$$-r_c(p, t, \alpha)G(q, p) + r_T(p, t, \alpha) = \dot{q}. \tag{15}$$

We recall Eq. (5), and see that

$$\dot{q} = F(q, p). \tag{16}$$

10. THE HARMONIC OSCILLATOR REVISITED

To illustrate the use of the above equations, let us return to the harmonic oscillator. Consider the governing equations

$$\dot{q} = \frac{p}{m}, \quad q(0) = w, \tag{1}$$

$$-\dot{p} = kq, \quad p(T) = c. \tag{2}$$

The relevant equations for r and τ are

$$r_T = \frac{c}{m} + krr_c \tag{3}$$

and

$$\tau_T = krr_c. \tag{4}$$

It is easily verified that a one-parameter family of solutions is given by

$$r(c, T, \alpha) = \frac{c}{\sqrt{km}} \tan\left[\sqrt{\frac{k}{m}} T + \alpha\right], \tag{5}$$

$$\tau(c, T, \alpha) = c \sec\left[\sqrt{\frac{k}{m}} T + \alpha\right]. \tag{6}$$

Consequently, the solution of the system of equations (1) and

(2) is

$$\beta = \tau(p, t, \alpha) = p \sec\left[\sqrt{\frac{k}{m}}\ t + \alpha\right],$$ (7)

and

$$q = r(p, t, \alpha)$$

$$= \frac{p}{\sqrt{km}}\ \tan\left[\sqrt{\frac{k}{m}}\ t + \alpha\right].$$ (8)

These expressions reduce to

$$p = \beta \cos\left[\sqrt{\frac{k}{m}}\ t + \alpha\right],$$ (9)

$$q = \frac{\beta}{\sqrt{km}}\ \sin\left[\sqrt{\frac{k}{m}}\ t + \alpha\right],$$ (10)

a form of the solution of the equations of the harmonic
oscillator.

11.   THIN BEAM THEORY

        As another example of the application of the appli-
cation of our ideas in mechanics, let us consider the determination of the equilibrium position of a thin beam.   In addition to its physical interest, this example will illustrate
the treatment of a fourth-order system of equations.

        The equation governing the position of a thin beam is

$$\frac{d^4u}{dt^4}(t) + k(t)u(t) = g(t), \quad 0 < t < L, \tag{1}$$

$$\frac{d^2u}{dt^2}(0) = 0, \quad \frac{d^3u}{dt^3}(0) = 0, \tag{2}$$

$$u(L) = \alpha, \quad \frac{du}{dt}(L) = \beta. \tag{3}$$

We wish to find a Cauchy system characterizing the function u. Let the linear operator A be given by

$$A[u] = \frac{d^4u}{dt^4} + k(t)u, \tag{4}$$

and consider the three associated linear problems for the functions v, w, z on the interval $0 \leq t \leq L$:

$$A[v] = g, \quad 0 < t < L, \tag{5}$$

$$\frac{d^2v}{dt^2}(0) = 0, \quad \frac{d^3v}{dt^3}(0) = 0, \tag{6}$$

$$v(L) = 0, \quad \frac{dv}{dt}(L) = 0; \tag{7}$$

$$A[w] = 0, \quad 0 < t < L, \tag{8}$$

$$\frac{d^2w}{dt^2}(0) = 0, \quad \frac{d^3w}{dt^3}(0) = 0, \tag{9}$$

$$w(L) = 1, \quad \frac{dw}{dt}(L) = 0; \tag{10}$$

$$A[z] = 0, \quad 0 < t < L, \tag{11}$$

$$\frac{d^2z}{dt^2}(0) = 0, \quad \frac{d^3z}{dt^3}(0) = 0, \tag{12}$$

$$z(L) = 0, \quad \frac{dz}{dt}(L) = 1. \tag{13}$$

The solution u can then be represented in the form

$$u(t) = v(t) + \alpha w(t) + \beta z(t), \quad 0 \le t \le L. \tag{14}$$

Viewing v, w, z explicitly as functions of L as well as t, and introducing the new functions m, n, p, q, r, s by the definitions

$$m(L) = \frac{d^2v}{dt^2}(L, L), \tag{15}$$

$$n(L) = \frac{d^2w}{dt^2}(L, L), \tag{16}$$

$$p(L) = \frac{d^2z}{dt^2}(L, L), \tag{17}$$

$$q(L) = \frac{d^3v}{dt^3}(L, L), \tag{18}$$

$$r(L) = \frac{d^3 w}{dt^3} (L, L),\tag{19}$$

$$s(L) = \frac{d^3 z}{dt^3} (L, L),\tag{20}$$

in the same manner as in Chapter 3, it is seen that the initial value system

$$m'(L) = q - mp, \quad m(0) = 0,\tag{21}$$

$$n'(L) = r - np, \quad n(0) = 0,\tag{22}$$

$$p'(L) = s - n - p^2, \quad p(0) = 0,\tag{23}$$

$$q'(L) = g - ms, \quad q(0) = 0,\tag{24}$$

$$r'(L) = -k - ns, \quad r(0) = 0,\tag{25}$$

$$s'(L) = -r - ps, \quad s(0) = 0,\tag{26}$$

is obtained. In the above, primes denotes $\frac{d}{dL}$. The equations satisfied by v, w, and z are

$$v'(t, L) = -mz, \quad v(t, t) = 0\tag{27}$$

$$w'(t, L) = -nz, \quad w(t, t) = 1,\tag{28}$$

$$z'(t, L) = -w - pz, \quad z(t, t) = 0, \quad L > t.\tag{29}$$

12. A SPECIFIC EXAMPLE

To test the numerical utility of the above equations, a specific example was solved. The problem chosen was that of a uniform beam of unit length, free at the left end and cantilevered at the right. The beam was loaded by a uniformly distributed load of unit magnitude, $g = 1$, and supported by an elastic foundation with a spring constant of $k = 1$. A reference solution for this problem was pre-computed by a segmenting technique. The results are shown in Table 2.

Table 2:

Deflection Calculations $(x10^{-1})$

t	Ref. Solution	Imbedding Solution
0.0	1.155238	1.155238
0.2	0.8485617	0.8485617
0.4	0.5502636	0.5502636
0.6	0.2821820	0.2821820
0.8	0.08126717	0.08126716
1.0	$-0.550387 \times 10^{-15}$	0.

13. OPTIMAL ESTIMATION AND FILTERING

An important problem in many areas of science and engineering is the estimation of a signal in the presence of

noise. It is well known in detection and estimation theory
that the solution of a Fredholm integral equation plays a
central role in the determination of the least-squares
estimate of a signal corrupted by additive white Gaussian
noise. In view of our earlier success with the treatment of
Fredholm integral equations, we suspect that the same basic
ideas will be applicable to the communication problem. Let
us now examine this conjecture in greater detail.

Consider the stochastic process

$$y(t) = z(t) + v(t), \qquad 0 \le t \le T \le \infty, \tag{1}$$

where v is a white Gaussian noise process with unit spectral
height such that

$$E[v(t)] = 0, \tag{2}$$

$$E[v(t)v(s)] = \delta(t - s), \qquad 0 \le t < s \le T, \tag{3}$$

where
$$E[z(t)v(s)] = 0, \qquad t < s, \tag{4}$$

$$E \equiv \text{expected value}.$$

For simplicity, we assume

$$E[z(t)] = 0, \qquad 0 \le t \le T, \tag{5}$$

and that v and z are scalars. Furthermore, assume that
the covariance K of the signal process z is given by

$$K(t, s) = E[z(t)z(s)], \quad 0 \le t, s \le T, \tag{6}$$

where K satisfies the following conditions

$$K(t, s) \text{ is continuous in } t \text{ and } s, \tag{7}$$

$$\int_0^T K(t, t) \, dt < \infty. \tag{8}$$

Introduce the quantity $\hat{z}(t)$ as the least-squares estimate of $z(t)$ given the observation $\{y(s), 0 \le s \le t\}$, $0 \le t \le T$. Standard results in filtering theory show that $\hat{z}$ may be written as

$$\hat{z}(t) = \int_0^t h(t, s)y(s) \, ds, \tag{9}$$

where the weighting function $h(t, s)$ satisfies the integral equation

$$h(t, s) = K(t, s) - \int_0^t K(\tau, s)h(t, \tau) \, d\tau,$$

$$0 \le s \le t \le T. \tag{10}$$

To apply our previous results to Eq. (10), let us introduce the additional hypothesis on K

$$K(t, s) = K(|t - s|) = \int_0^1 e^{-|t-s|\alpha} w(\alpha) \, d\alpha. \tag{11}$$

This form is of considerable interest in filtering corresponding to a generalized form of a Butterworth process.

14. INVARIANT IMBEDDING AND REAL-TIME FILTERING

Let us now indicate how the results and techniques of Chapter 4 allow us to develop a real-time filter for the problem stated above.

Let the function J satisfy the stated integral equation

$$J(s, t, \alpha) = e^{-(t-s)\alpha} - \int_0^t K(|\tau - s|)J(\tau, t, \alpha) \, d\tau,$$

$$0 \leq s \leq t \leq T, \quad 0 \leq \alpha \leq 1. \tag{1}$$

In terms of J, we may express the function h as

$$h(t, s) = \int_0^1 J(s, t, \alpha')w(\alpha') \, d\alpha'. \tag{2}$$

Following exactly the same sequence of steps as given in Chapter 4, the Cauchy system for J is obtained as

$$X_t(t, \alpha) = -Y(t, \alpha)\int_0^1 Y(t, \alpha')w(\alpha') \, d\alpha', \tag{3}$$

$$Y_t(t, \alpha) = -\alpha Y(t, \alpha) -$$

$$X(t, \alpha) \int_0^1 Y(t, \alpha')w(\alpha') \, d\alpha', \tag{4}$$

$$J_t(s, t, \alpha) = -\alpha J(s, t, \alpha) -$$

$$X(t, \alpha) \int_0^1 J(s, t, \alpha')w(\alpha') \, d\alpha',$$

$$0 \le s \le t \le T, \qquad 0 \le \alpha \le 1. \tag{5}$$

The initial conditions are

$$X(0, \alpha) = 1, \tag{6}$$

$$Y(0, \alpha) = 1, \tag{7}$$

$$J(s, s, \alpha) = X(s, \alpha). \tag{8}$$

At this point we could compute h through Eq. (2). However, let us follow another path and obtain an equation for $\hat{z}$ which involves only X, Y, and another auxiliary function of t and α. By eliminating J in this manner, signifi-cant computational savings can be anticipated.

To this end, recall the equation for $\hat{z}$,

$$\hat{z}(t) = \int_0^t h(t, s) \, y(s) \, ds, \quad t > 0. \tag{9}$$

Making use of Eq. (2), we see that

$$\hat{z}(t) = \int_0^t \int_0^1 J(s, t, \alpha')w(\alpha') \, d\alpha' \, y(s) \, ds. \qquad (10)$$

Interchanging the order of integration in Eq. (10) gives

$$\hat{z}(t) = \int_0^1 \int_0^t J(s, t, \alpha')y(s) \, ds \, w(\alpha') \, d\alpha'. \qquad (11)$$

Introduce the new function $L(t, \alpha)$ as

$$L(t, \alpha) = \int_0^t J(s, t, \alpha)y(s) \, ds, \quad t > 0,$$

$$0 \le \alpha \le 1. \qquad (12)$$

Differentiate L with respect to t to obtain

$$L_t(t, \alpha) = J(t, t, \alpha)y(t) + \int_0^t J_t(s, t, \alpha)y(s) \, ds,$$

$$t > 0, \quad 0 \le \alpha \le 1. \qquad (13)$$

Recalling Eqs. (5), (8) and (12), we observe that Eq. (13) may be written

$$L_t(t, \alpha) = X(t, \alpha)y(t) - \alpha L(t, \alpha) -$$

$$X(t, \alpha)\int_0^1 L(t, \alpha')w(\alpha') \, d\alpha',$$

$$= -\alpha L(t, \alpha) + X(t, \alpha)\{y(t) -$$

$$\int_0^1 L(t, \alpha')w(\alpha')\ d\alpha'\}, \quad 0 \le s \le t \le T,$$

$$0 \le \alpha \le 1. \quad (14)$$

The initial condition is given by Eq. (12) as

$$L(0, \alpha) = 0. \tag{15}$$

Knowledge of L then allows $\hat{z}$ to be written as

$$\hat{z}(t) = \int_0^1 L(t, \alpha')w(\alpha')\ d\alpha'. \tag{16}$$

The equations for X, Y, and L, together with relation (16) for $\hat{z}$, comprise an initial value system suitable
for obtaining the least squares estimate of the signal
process in the noisy environment described above. Since the
observed function y appears in the system as a forcing
term in Eq. (14), the system is ideally suited for use in a
real-time situation. Similar results are easy to derive for
the multidimensional case as well as for situations in which
the covariance kernel is not of the form assumed in Eq.
(13.11). We defer discussion of these important topics to a
forthcoming volume.

15. SOME ASPECTS OF NONLINEAR SMOOTHING

A far more interesting and, of course, more difficult
class of filtering problems arises when the system dynamics
and/or observations are nonlinear functions of the state.
Depending upon the instant at which estimates are desired,
we are faced with a prediction, filtering, or smoothing problem as the estimation is for future, current, or past time,
respectively.

We wish to find a sequential solution for the nonlinear smoothing problem with a least-squares cost function.
In the smoothing problem, primary interest is focused on
obtaining an estimate of states and parameters of the system
at a finite set of fixed instants of time which are contained
in the interval of observation.

Problems of this nature frequently arise in astronomy
where one wishes to update estimates of initial conditions
as additional observations become available and in trajectory
analysis problems in which one may wish to improve estimates
at "epochs". Typically, these epochs are associated with
mid-course connection times, instants at which photographs
are taken, and so forth.

The process to be estimated is described by

$$\dot{x}(t) = g(t, x) + \text{dynamical error} \tag{1}$$

and observations on the process are made in the form

$$y(t) = h(t, x) + \text{observational error} \qquad (2)$$

over the interval $0 \le t \le T$. It is required to estimate the state of the process at a fixed instant t_1, $0 \le t_1 \le T$, based on knowledge of y in the interval $0 \le t \le T$. The extension to the case where there are several times $\{t_i\}$ at which estimates are desired is straightforward and will be left as a simple exercise for the reader.

The dynamical and observational error in Eqs. (1) and (2) account for imprecise knowledge of the right-hand sides, as well as for any stochastic disturbances.

The optimal estimate $x^*(t)$ minimizes the least-squares criteria

$$I(z) = \int_0^T [k_1 \{y(t) - h(t, z)\}^2 +$$

$$k_2 \{z - g(t, z)\}^2] \, dt + k_3 \{z(0) - m_0\}^2 \qquad (3)$$

where k_1 and k_2 are non-negative weighting functions and k_3 is a non-negative constant. The optimal estimate at time $t_1 \le T$ is then given by $x^*(t_1)$.

The determination of the optimal estimate is clearly equivalent to the following optimal control problem.

Determine the optimal control $u(t)$ and the optimal trajectory $x(t)$, $0 \le t \le T$, to minimize

$$I(u) = \int_0^T [k_1\{y(t) - h(t, x)\}^2 +$$

$$k_2 u^2(t)] \ dt + k_3[x(0) - m_0]^2 \quad (4)$$

subject to the constraint

$$\dot{x} = g(x, t) + u$$

with $x(0)$ and $x(T)$ being free.

16.   THE SEQUENTIAL SMOOTHER

The simplest route to an initial value problem for $x^*(t_1)$ is through the Euler equation associated with the functional (15.4) and constraint (15.5). This equation is

$$\dot{x} = g(t, x) - \frac{\lambda}{2k_2} , \quad \lambda(0) - 2k_3[x(0) - m_0] = 0 \quad (1)$$

$$\dot{\lambda} = 2k_1\{y - h(t, x)\}h_x(t, x) - g_x(t, x)\lambda,$$

$$\lambda(T) = 0. \quad (2)$$

For sequential estimation, we desire a set of equations which give $x(t_1)$ as a function of T. To obtain these relations, consider the problem

$$\dot{x} = \alpha(t, x, \lambda), \quad \lambda(0) - 2k_3[x(0) - m_0] = 0 \tag{3}$$

$$\dot{\lambda} = \beta(t, x, \lambda), \quad \lambda(T) = c, \quad T \geq 0, \quad |c| < \infty. \tag{4}$$

Adopting the notation used before for the value of $x(t)$ at t_1, namely $x(t_1, T, c)$, the arguments given in Chapter 3 give the Cauchy problem

$$x_T(t_1, T, c) + \beta(T, r(c, T), c)x_c(t_1, T, c) = 0 \tag{5}$$

$$x(t_1, t_1, c) = r(t_1, c), \tag{6}$$

$$r_T(T, c) + \beta(T, r(T, c), c)r_c(T, c) =$$

$$\alpha(T, r(T, c), c), \tag{7}$$

$$r(0, c) = \frac{c}{2k_3} + m_0. \tag{8}$$

In spite of its foreboding appearance, the system of equations (5)-(8) can be handled by finite differences, perturbation techniques, or other means. The solution of interest, of course, is $x(t_1, T, 0)$.

17. LINEAR SYSTEMS

Of special interest is the case when the functions g

and h are linear in x. In this situation, the equations

given above reduce to ordinary differential equations via

separation of variables.

If

$$g(t, x) = g(t)x + f(t),$$ (1)

$$h(t, x) = h(t)x,$$ (2)

then

$$\alpha(t, x, \lambda) = g(t)x + f(t) \frac{\lambda}{2k_2},$$ (3)

$$\lambda(t, x, \lambda) = 2k_1[y(t) - h(t)x]h(t) - g(t)\lambda.$$ (4)

Writing

$$r(T, c) = p(T)c + q(T),$$ (5)

and equating like powers of c in Eq. (16.7), we obtain

$$\dot{p}(T) = 2k_1 h^2(T)p^2(T) + 2g(T)p(T) - \frac{1}{2k_2},$$ (6)

$$\dot{q}(T) = g(T)q(T) + f(T) -$$
$$\qquad\qquad 2k_1 h(T)p(T)[y(T) - h(T)q(T)],$$ (7)

with the initial conditons

$$p(0) = \frac{1}{2k_3},$$ (8)

$$q(0) = m_0. \tag{9}$$

Similarly, if we write

$$x(t_1, T, c) = S(t_1, T)c + u(t_1, T) \tag{10}$$

and equate like powers of c in Eq. (16.5) the result is

$$S_T(t_1, T) = g(T)S(t_1, T) + 2k_1 h^2(T)p(T), \tag{11}$$

$$u_T(t_1, T) = -2k_1 h(T)S(t_1, T)[y(T) - h(T)q(T)], \tag{12}$$

$$S(t_1, t_1) = p(t_1), \tag{13}$$

$$u(t_1, t_1) = q(t_1). \tag{14}$$

Since the optimal estimate at time t_1 is $x(t_1, T, 0)$, it follows that $u(t_1, T)$ is the estimate we seek. Thus, Eqs. (6)-(14) are the sequential smoothing equations for the linear system.

18. INTEGRO-DIFFERENTIAL EQUATIONS AND NONLOCAL WAVE
     INTERACTION

In the study of electromagnetic fields in metallic slabs and semi-conductors of finite thickness, certain integro-differential equations occur for the electric field.   These equations have the form

$$\frac{d^2e}{dt^2}(t) + Ae(t) = \int_0^L k(|t - t'|)e(t') \, dt',$$

$$0 \le t \le L, \tag{1}$$

where A is a constant. In addition, boundary conditions
are specified at $t = 0$ and $t = L$.

In the foregoing pages, we have seen that various
two-point boundary value problems, integral equations, and
variational problems can be transformed into initial value
systems. The remaining sections will show that another large
class of functional equations, integro-differential equations,
subject to boundary conditions can be transformed into Cauchy
systems. Such equations arise in modern physics where non-
local interactions must be considered.

To illustrate the ideas involved, consider determin-
ing the function $u(t)$, $0 \le t \le c$, which is a solution of the
integro-differential equation

$$\ddot{u}(t) + Au(t) = \int_0^c k(|t - y|)u(y) \, dy, \tag{2}$$

with boundary conditions

$$\dot{u}(0) = 0, \; u(c) = 1. \tag{3}$$

As before, we assume the kernel k has the form

$$k(r) = \int_a^b e^{-r/z'} w(z') \, dz', \quad r > 0. \tag{4}$$

19. STATEMENT OF INITIAL VALUE SYSTEM

Since the derivation of the initial value system is lengthy, we state the final result before entering into the technical details. Let the functions u and v be solutions of the linear two-point boundary value problem

$$\dot{u}(t, x) = v(t, x), \tag{1}$$

$$\dot{v}(t, x) + Au(t, x) = \int_0^x k(|t - y|)u(y, x) \, dy,$$

$$0 \le t \le x \le c, \tag{2}$$

$$v(0, x) = 0, \tag{3}$$

$$u(x, x) = 1. \tag{4}$$

The initial value problem, suitable for determining v and u, is comprised of the eight functions e, r, α, R, u, v, J, and M which satisfy the equations

$$e_x(x, s) = 1 - \frac{1}{s} \, e(x, s) = R(x)e(x, s) +$$

$$\int_a^b r(s, z', x)w(z') \, dz', \tag{5}$$

$$r_x(s, z, x) = -(\frac{1}{s} + \frac{1}{z})r(s, z, x) -$$

$$\alpha(x, z)e(x, s), \tag{6}$$

$$\alpha_x(x, z) = 1 + \int_a^b r(z', z, x)w(z') \, dz' -$$

$$\alpha(x, z)[\frac{1}{z} + R^2(x)], \tag{7}$$

$$R_x(x) = -A + \int_a^b [e(x, z') +$$

$$\alpha(x, z')]w(z') \, dz' - R^2(x), \tag{8}$$

$$u_x(t, x) = -R(x)u(t, x) +$$

$$\int_a^b J(t, w, z')w(z') \, dz', \tag{9}$$

$$v_x(t, x) = -R(x)v(t, x) +$$

$$\int_a^b M(t, x, z')w(z') \, dz', \tag{10}$$

$$J_x(t, x, z) = -\frac{1}{z} J(t, x, z) - \alpha(x, z)u(t, x), \tag{11}$$

$$M_x(t, x, z) = -\frac{1}{z} M(t, x, z) -$$

$$\alpha(x, z)v(t, x), \quad 0 \le t \le x \le c, \quad a \le s, z \le b. \quad (12)$$

The initial conditions are

$$e(0, s) = 0, \tag{13}$$

$$r(s, z, 0) = 0, \tag{14}$$

$$\alpha(0, z) = 0, \tag{15}$$

$$R(0) = 0, \tag{16}$$

$$u(t, t) = 1, \tag{17}$$

$$v(t, t) = R(t), \tag{18}$$

$$J(t, t, z) = 0, \tag{19}$$

$$M(t, t, z) = \alpha(t, z). \tag{20}$$

Equations (5)-(20) may now be used in the same manner as before to produce the functions of interest, u and v.

20. DERIVATION OF THE CAUCHY SYSTEM

We begin our derivation of the above relations by differentiating Eqs. (19.1)-(19.4) with respect to x. This gives

$$\dot{u}_x(t, x) = v_x(t, x), \tag{1}$$

$$\dot{v}_x(t, x) + Au_x(t, x) = k(x - t)u(x, x) +$$

$$\int_0^x k(|t - y|)u_x(y, x) \, dy, \qquad 0 \leq t \leq x, \quad (2)$$

$$v_x(0, x) = 0, \tag{3}$$

$$\dot{u}(x, x) + u_x(x, x) = 0. \tag{4}$$

In Eq. (4) dot denotes differentiation with respect to the
first argument while the subscript x represents differentiation with respect to the second.

Introduce the functions Φ and Ψ as the solutions
of the inhomogeneous integro-differential equations

$$\dot{\Phi}(t, x) = \Psi(t, x), \tag{5}$$

$$\dot{\Psi}(t, x) + A\Phi(t, x) = k(x - t) +$$

$$\int_0^x k(|t - y|)\Phi(y, x) \, dy, \qquad 0 \leq t \leq x, \quad (6)$$

with the homogeneous houndary conditions

$$\Psi(0, x) = 0, \tag{7}$$

$$\Phi(0, x) = 0. \tag{8}$$

Regarding Eqs. (1)-(4) as an inhomogeneous system of integro-

differential equations for u_x and v_x subject to inhomogeneous boundary conditions, the principle of superposition
provides the solutions

$$u_x(t, x) = -\dot{u}(x, x)u(t, x) +$$

$$u(x, x)\Phi(t, x), \quad (9)$$

$$v_x(t, x) = -\dot{u}(x, x)v(t, x) +$$

$$u(x, x)\Psi(t, x), \quad x \geq t. \quad (10)$$

The boundary condition in Eq. (19.4) disposes of $u(x, x)$
(=1). The functions Φ, Ψ, and $-\dot{u}(x, x)$ will now be considered.

Let the new functions J and M be defined as the
solutions of the system

$$\dot{J}(t, x, z) = M(t, x, z), \quad (11)$$

$$\dot{M}(t, x, z) + AJ(t, x, z) = e^{-(x-t)/z} +$$

$$\int_0^x k(|t - y|)J(y, x, z)\,dy,$$

$$0 \leq t \leq x, \; a \leq z \leq b, \quad (12)$$

$$M(0, x, z) = 0, \quad (13)$$

$$J(x, x, z) = 0. \quad (14)$$

In view of the representation for the kernel k in Eq.
(18.4) and Eqs. (5)-(8), it is clear that we may write

$$\Phi(t, x) = \int_a^b J(t, x, z')w(z') \, dz',$$
(15)

$$\Psi(t, x) = \int_a^b M(t, x, z')w(z') \, dz', \quad 0 \le t \le z.$$
(16)

We now shift our attention to the determination of J and
M.

Through differentiation with respect to x, Eqs. (11)-
(14) become

$$\dot{J}_x(t, x, z) = M_x(t, x, z),$$
(17)

$$\dot{M}_x(t, x, z) + AJ_x(t, x, z) =$$

$$- \frac{1}{z} e^{-(x-t)/z} + k(x - t)J(x, x, z) +$$

$$\int_0^x k(|t - y|)J_x(y, x, z) \, dy,$$
(18)

$$M_x(0, x, z) = 0,$$
(19)

$$\dot{J}(x, x, z) + J_x(x, x, z) = 0.$$
(20)

Since

$$J(x, x, z) = 0$$
(21)

and

$$-\dot{J}(x, x, z) = -M(x, x, z),$$
(22)

it is seen that

$$J_x(t, x, z) = -\frac{1}{z} J(t, x, z) -$$

$$M(x, x, z)u(t, x),$$
(23)

$$M_x(t, x, z) = -\frac{1}{z} M(t, x, z) -$$

$$M(x, x, z)v(t, x), \quad x > t, \quad a \le z \le b.$$
(24)

Define the function α to be

$$\alpha(x, z) = M(x, x, z).$$
(25)

It follows from Eqs. (12) and (24) that

$$\alpha_x(x, z) = \dot{M}(x, x, z) + M_x(x, x, z) =$$

$$1 + \int_0^x k(x - y)J(y, x, z) \, dy -$$

$$\alpha(x, z)[\frac{1}{z} + v(x, x)].$$
(26)

It is now convenient to introduce the additional terminology
that

$$r(s, z, x) = \int_0^x e^{-(x-y)/s} J(y, x, z) \, dy,$$

$$a \le s, \quad z \le b, \quad x \ge 0,$$
(27)

and

$$R(x) = v(x, x), \quad x \geq 0. \tag{28}$$

Again recalling the representation for the kernel k in

Eq. (18.4), we write

$$\alpha_x(x, z) = 1 + \int_0^x \int_a^b e^{-(x-y)/z'} w(z') dz' J(y, x, z) \, dy$$

$$- \alpha(x, z)[\frac{1}{z} + R(x)]. \tag{29}$$

This becomes, finally,

$$\alpha_x(x, z) = 1 + \int_a^b r(z', z, x) w(z') \, dz' -$$

$$\alpha(x, z)[\frac{1}{z} + R(x)]. \tag{30}$$

Now we turn to the function r. Through differentia-

tion of both sides of Eq. (27) with respect to x, we find

that

$$r_x(s, z, x) = J(x, x, z) - \frac{1}{s} r(s, z, x) +$$

$$\int_0^x e^{-(x-y)/s}[-\frac{1}{z} J(y, x, z) -$$

$$\alpha(x, z) u(y, x)] \, dy. \tag{31}$$

Upon simplification, this last equation becomes

$$r_x(s, z, x) = -\left(\frac{1}{z} + \frac{1}{s}\right)r(s, z, x) -$$

$$\alpha(x, z)e(x, s), \quad (32)$$

where

$$e(x, s) = \int_0^x e^{-(x-y)/s}u(y, x)\, dy,$$

$$x \geq 0, \quad a \leq s \leq b. \quad (33)$$

Let us now consider the function   e.   From Eq. (33), we see that

$$e_x(x, s) = u(x, x) - \frac{1}{s}e(x, s) +$$

$$\int_0^x e^{-(x-y)/s}[-\dot{u}(x, x)u(y, x) +$$

$$u(x, x)\Phi(y, x)]\, dy. \quad (34)$$

From this equation it is seen that

$$e_x(x, s) = 1 - \frac{1}{s}e(x, s) -$$

$$R(x)e(x, s) + \int_0^x e^{-(x-y)/s}\Phi(y, x)\, dy. \quad (35)$$

Note that

$$\dot{u}(x, x) = v(x, x) = R(x). \quad (36)$$

The integral in Eq. (35) is transformed by employing Eq. (15).

It becomes

$$\int_0^x e^{-(x-y)/s} \int_a^b J(y, x, z')w(z') \, dz' \, dy =$$

$$\int_a^b w(z') \, dz' \int_0^x e^{-(x-y)/s} J(y, x, z') \, dy, \qquad (37)$$

$$\int_0^x e^{-(x-y)/s} \Phi(y, x) \, dy = \int_a^b r(s, z', x)w(z') \, dz'. \qquad (38)$$

The result is that the differential equation for e is

$$e_x(x, s) = 1 - \frac{1}{s} e(x, s) - R(x)e(x, s) +$$

$$\int_a^b r(s, z', x)w(z') \, dz', \qquad a \le s \le b. \quad (39)$$

Lastly, we turn our attention to the function R, defined earlier to be

$$R(x) = v(x, x). \qquad (40)$$

Differentiation shows that

$$R_x(x) = \dot{v}(x, x) + v_x(x, x)$$

$$= -Au(x, x) + \int_0^x k(x - y)u(y, x) \, dy -$$

$$\dot{u}(x, x)v(x, x) + \Psi(x, x)u(x, x)$$

$$= -A + \int_0^x \int_a^b e^{-(x-y)/z'} w(z') \, dz' \, u(y, x) \, dy -$$

$$R^2(x) + \int_a^b M(x, x, z')w(z') \, dz'$$

$$= -A + \int_a^b e(x, z')w(z') \, dz' - R^2(x) +$$

$$\int_a^b \alpha(x, z')w(z')dz'. \quad (41)$$

The differential equations for the functions R, e, α, and r have now been obtained. They are Eqs. (41), (39), (30), and (32). From the definitions, it is seen that the initial conditions at x = 0 are all zero for all a $\leq$ z, s $\leq$ b.

Let t be a fixed non-negative number in the interval [0, x]. Then, the differential equations for the functions J, M, u, and v are given in Eqs. (23), (24), (9), and (10). This completes the derivation of the Cauchy system.

CHAPTER SIX

NOTES AND REFERENCES

§2.   Detailed analysis of the physical aspects of the

source function are provided in

Sobolev, V. V., A Treatise on Radiative Transfer,

Van Nostrand Co., New York, 1963.

and in

Chandrasekhar, S., Radiative Tranfer, Dover Co.,

New York, 1960.

§3.   Extensive numerical results and graphs of the source

function are in

Bellman, R., H. Kagiwada, and R. Kalaba, "Numerical

Results for the Auxiliary Equations of Radiative

Transfer," J. Quant. Spec. Rad. Trans., 6, 1966,

219-310.

§4.   Reflection and Transmission functions are tabulated

in the book

Bellman, R., R. Kalaba, and M. Prestrud, Invariant

Imbedding and Radiative Transfer in Slabs of Finite

Thickness, American Elsevier Co., New York, 1963.

§5-7. We follow the material in

Casti, J., R. Kalaba, and S. Ueno, "Reflection and Transmission Functions for Finite Isotropically Scattering Atmospheres with Specular Reflectors," J. Quant. Spec. Rad. Trans., 9, 1969, 537-552.

§8-10. These results were first given in

Bellman, R. and R. Kalaba, "A Note on Hamilton's Equations and Invariant Imbedding," Quart. Appl. Math., 21, 1963, 166-168.

Bellman, R., H. Kagiwada, and R. Kalaba, "Invariant Imbedding and Nonvariational Principles in Analytical Dynamics," Int'l. J. Nonlinear Mech., 1, 1966, 51-55.

§11-12. Further extensions and more extensive numerical results are provided in the papers

Alspaugh, D., H. Kagiwada, and R. Kalaba, "Dynamic Programming, Invariant Imbedding, and Thin Beam Theory," the RAND Corporation, RM-5706-PR, Aug. 1968.

Alspaugh, D., H. Kagiwada, and R. Kalaba, "Application of Invariant Imbedding to the Eigenvalue Problems for Buckling of Columns," The RAND Corp., RM-5954-PR, Aug. 1969.

§13-17. For a more detailed discussion of the filtering

problem see

Casti, J., R. Kalaba, and V.K. Murthy, "A New Ini-

tial Value Method for On-Line Filtering and Estima-

tion," IEEE Trans. Info. Th., July 1972, 515-517.

The results for the nonlinear case are in

Kagiwada, H., R. Kalaba, A. Schumitzky, and R.

Sridhar, "Invariant Imbedding and Sequential

Interpolating Filters for Nonlinear Processes,"

The RAND Corp., RM-5507-PR, Nov. 1967.

A related discussion together with several numerical

experiments is presented in

Bellman, R., H. Kagiwada, R. Kalaba, and R.

Sridhar, "Invariant Imbedding and Nonlinear Filter-

ing Theory," J. Astro. Sci., 13, 1966, 110-115.

It would be of interest to compare the results for

the special case of linear systems with the standard

work in this field

Kalman, R. and R. Bucy, "New Results in Linear

Filtering and Prediction Theory," J. Basic Engr.,

ASME Series D, 83, 161, 95-108.

§18.   Applications of these integro-differential equations

in solid-state plasmas, electron-electron interactions, the anomalous skin effect, and helicon
propagatious near doppler-shifted cyclotron reson-
ance are found in the decisive papers

Baraff, G., J. Math. Phys., 9, 1968, 372,

Baraff, G., Phys. Rev., 167, 1968, 625,

Baraff, G., Phys. Rev., 178, 1969, 1155,

Juras, G., "Line Shapes in the Radio Frequency Size
Effect of Metals," Phys. Rev., 187, 1969, 784.

§19-20. Our development was first presented in

Kalaba, R., "Boundary Value Problems for the
Integro-Differential Equations of Nonlocal Wave
Interaction-I," J. Math. Phys., 11, 1970, 1999-
2004.

APPENDIX A

GENERAL COMPUTER PROGRAM FOR
INTEGRAL EQUATIONS WITH DISPLACEMENT KERNELS

1. DESCRIPTION

The Program assumes that the kernel $k(t, y)$ is expressed in the form given in Eq. (2.2) of Chapter IV when $z = [0, 1]$. The user must supply the subroutine KERNEL which computes the function $w(z_i) \cdot WT_i$, where the z_i's are the gaussian quadrature points on the interval $[0, 1]$, $i = 1, 2, \ldots, $ NQUAD (the order of the quadrature chosen by the user), and the WT_i's are the corresponding quadrature weights. The values $w(z_i) \cdot WT_i$ are stored in the vector AKERN(I). The user must also supply a subroutine ESUM1. Its purpose is to compute the forcing function $g(t)$. The argument of ESUM1 is the current value of the independent variable x. The result of the calculation is stored in variable location E1SUM.

The entry point for the program is the statement CALL INTEQ (NQUAD, NIPTS, PTS, ALEN(K), TDEL, OUTPT, T, K, INDM1) where the following quantities are to be supplied by the user in his calling program:

NQUAD — order of the quadrature scheme, an integer

from 3 to 15.

285

NIPTS   — an integer from 1 to 76 indicating the
         number of interval points at which a solu-
         tion is desired.

PTS   — a vector of dimension 76 in which the
         desired interval points are stored in
         ascending order.

ALEN(K) — a real variable indicating the total
         interval length for which solution is
         desired.

TDEL   — a real variable indicating the integration
         step size to be used. (It must be _greater_
         than 0.001.)

OUTPT   — a vector of dimension 76 in which the solu-
         tion values at the interval points are
         stored.   (The values in OUTPT are in 1-1
         correspondence with the interval points of
         vector PTS.)

T   — a storage vector of dimension 12610.

NTL   — number of interval lengths at which the
         solution is to be output.

The user should note that the interval points in PTS must be
integral multiples of the step size TDEL.

At the conclusion of a run with the user-supplied
step size TDEL, the program will halve the step size and

resolve the program as a check on the appropriateness of

the original step size if the variable NCHECK is set equal

to some nonzero integer. In this case the user may obtain

the results from the check run by changing the main program

to write the vector TEST instead of the vector OUTPT.

Should the user desire the values of the auxiliary

quantities X_i, Y_i, e_i, J_i, or u_i, they are stored in the

T vector as follows:

$$X_i \quad - \quad T(1) - T(NQUAD)$$

$$Y_i \quad - \quad T(NQUAD+1) - T(2*NQUAD)$$

$$e_i \quad - \quad T(2*NQUAD+1) - T(3*NQUAD)$$

$$J_{1,i} \quad - \quad T(3*NQUAD+1) - T(4*NQUAD)$$

$$u_1 \quad - \quad T(4*NQUAD+1)$$

$$J_{2,i} \quad - \quad T(4*NQUAD+2) - T(5*NQUAD+1)$$

$$u_2 \quad - \quad T(5*NQUAD+2)$$

$$\begin{array}{cc} \cdot & \cdot \\ \cdot & \cdot \\ \cdot & \cdot \end{array}$$

$i = 1, 2, \ldots, NQUAD$

Should the user desire the solution at more than the

76 points allowed for in the listing given in the Appendix,

the following steps should be taken:

1. Increase the dimension of the vectors UPRI, PTS, and OUTPT to the desired number of solution points.

2. Increase the dimension of vector T from 12610 to 450 + 160 NIPTS.

3. Increase the dimension of array AJPRI from (76,15) to (NIPTS,15).

4. In subroutines INT1 and INT2 change the dimension of the vector DUMMY from 1310 to 95 + 160 · NIPTS.

2. ARRANGEMENT OF INPUT DATA

The program calls for the input data to be submitted in the following order and form:

			Format
Card 1 NQUAD, NIPTS	:		2I5
Card 2 NIL	:		I5
Card 3 to Card (NIPTS+1)	:	PTS	F10.3
Card (NIPTS+2)	:	TDEL	F10.3
Card (NIPTS+3) - (NIPTS+3+NIL)	:	ALEN	F10.3
Card (NIPTS+NIL+4)	:	NCHECK	I5

The integration routine used for this program is a fourth-order Adams-Moulton predictor corrector utilizing a fourth-order Runge-Kutta method to start the process.

```
//C2600#04 JOB (6547,120,152),'IMBEDDING',CLASS=B
//STEP1 EXEC FORTCLG,REGION.GO=200K
//FORT.SYSIN DO #
      IMPLICIT REAL#8(A-H,O-Z)
      DIMENSION PTS(76),OUTPT(76),TEST(76),T(12610),ALEN(10)
      DO 1 I=1,30
      OUTPT(I)=0.
    1 TEST(I)=0.
      READ(5,100)NQUAD,NIPTS
      READ(5,100)NIL
      READ(5,101)(PTS(I),I=1,NIPTS)
      READ(5,101)TDEL
      READ(5,101)(ALEN(I),I=1,NIL)
      READ(5,100)NCHECK
      DO 20 K=1,NIL
      WRITE(6,102)ALEN(K)
    8 CALL INTEQ(NQUAD,NIPTS,PTS,ALEN(K),TDEL,OUTPT,T,K,
      INDM1)
      IF(NCHECK.EQ.O)GO TO 9
      DO 5 I=1,30
      IF(DABS(TEST(I)-OUTPT(I)).LE.1.D-4)GO TO 5
      GO TO 6
    5 CONTINUE
      GO TO 9
    6 DO 7 I=1,30
    7 TEST(I)=OUTPT(I)
      TDEL=TDEL/2.
      GO TO 8
    9 DO 10 I=1,INDM1
   10 WRITE(6,103)PTS(I),OUTPT(I)
      WRITE(6,104)TDEL
   20 CONTINUE
      CALL EXIT
  100 FORMAT(2I5)
  101 FORMAT(F10.3)
  102 FORMAT(1H1,28HSOLUTION FOR INTERVAL LENGTH,D10.3/1H,
      5X,1HT,10X,4H,
     1U(T))
  103 FORMAT(1H,F6.3,D14.7)
  104 FORMAT(1H0,17HFINAL STEP SIZE=.D13.6)
      END
      SUBROUTINE INTEQ(NQ,NIPTS,PTS,ALENG,H,OUTPT,T,KK,INDM1)
C
C
C     -NOTATION-
C
C     R-VECTOR OF ROOTS
C     W-VECTOR OF WEIGHTS
```

```
C NQUAD-NUMBER OF QUADRATURE POINTS
C AKERN-VECTOR OF W(Z) FUNCTIONS
C  XSUM-
C  YSUM- SUMS ASSOCIATED WITH X,Y,AND E FUNCTIONS
C E1SUM-
C E2SUM-
C  XPRI-
C  YPRI- VECTORS ASSOCIATED WITH DERIVATIVES OF X,Y,E,J AND
C    U FUNCTIONS
C  EPRI-
C AJPRI-
C  UPRI-
C    PTS-VECTOR OF INTERNAL POINTS AT WHICH A SOLUTION IS
C     DESIRED
C  ALEN-VECTOR OF INTERVAL LENGTHS FOR WHICH SOLUTIONS ARE
C     DESIRED
C   NIL-VARIABLE INDICATING NUMBER OF INTERVAL LENGTHS IN
C     VECTOR ALEN
C NIPTS-VARIABLE INDICATING NUMBER OF INTERNAL POINTS TO BE
C     CONSIDERED
C
      IMPLICIT REAL*8(A-H,O-Z)
      COMMON                    R(15),W(15),AKERN(15),XSUM,YSUM,
     ElSUM,E2SUM
     1,XPRI(15),YPRI(15),EPRI(15),AJPRI(76,15),UPRI(76)
      COMMON AUX(4),NAUX(11),NQUAD
      DIMENSION PTS(76),OUTPT(76),T(12610)
      IF(KK.NE.1)GO TO 35
      NQUAD=NQ
      CALL RTSWTS
      CALL KERNEL
      DO 5 I=1,5250
    5 T(I)=0.
C
C
C
      X=0.
C
C
C     N=NUMBER OF DIFFERENTIAL EQUATIONS
C
      N=3*NQUAD
      N1=2*NQUAD
      DO 10 I=1,N1
   10 T(I)=1.
      IND=1
      CALL INT1(T,N,X,H)
```

```
C
C     IF INTEGRATION HAS REACHED AN INTERNAL POINT,ADJOIN NEW
C       EQUATIONS
C
   11 IF(DABS(X-PTS(IND)).LE.L.D-3)GO TO 14
   35 IF(DABS(X-ALENG).LE.1.D-3)GO TO 37
   13 CALL INT2(T,N,X,H)
      GO TO 11
   14 N=N+NQUAD+1
      NEND=NQUAD
      DO 20 I=1,NEND
      K=N+I-NQUAD-1
   20 T(K)=T(I)
      CALL ESUM1(X)
      CALL ESUM2(T,N,X,H)
      T(K+1)=E1SUM+E2SUM
      CALL INT1(T,N,X,H)
      IND=IND+1
      GO TO 35
   37 INDM1=IND-1
      DO 40 I=1,INDM1
      IZ=(I+3)*(NQUAD+1)-3
   40 OUTPT(I)=T(IZ)
      RETURN
      END
      SUBROUTINE DAUX(T,N,X,H)
      IMPLICIT REAL*8(A-H,O-Z)
      COMMON                    R(15),W(15),AKERN(15),XSUM,YSUM,
       E1SUM,E2SUM
     1,XPRI(15),YPRI(15),EPRI(15),AJPRI(76,15),UPRI(76)
      COMMON AUX(4),NAUX(11),NQUAD
      DIMENSION T(12610)
      CALL XSUMS(T,N)
      CALL YSUMS
      CALL ESUM1(X)
      CALL ESUM2(T,N,X,H)
      CALL PRIMES(T,N,X,H)
      IF(N.EQ.(3*NQUAD))GO TO 5
      GO TO 15
    5 DO 10 I=1,NQUAD
      K=I+3*(NQUAD+1)-3
      T(K)=XPRI(I)
      KK=K+NQUAD
      T(KK)=YPRI(I)
      KKK=KK+NQUAD
   10 T(KKK)=EPRI(I)
      RETURN
   15 NJ=(N-3*NQUAD/(NQUAD+1)
      NSTRT= NQUAD*(3+NJ)+NJ
```

```
        NEND=3*NQUAD+N+3
        DO 20 I=1,NQUAD
        IX=NSTRT+I
        T(IX)=XPRI(I)
        IY=IX+NQUAD
        T(IY)=YPRI(I)
        EI=IY+NQUAD
  20    T(IE)=EPRI(I)
        JNO=1
  25    DO 30 I=1,NQUAD
        IJ=(JNO-1)+(NQUAD+1)+NEND+I-3
  30    T(IJ)=AJPRI(JNO,1)
        T(IJ+1)=UPRI(JNO)
        JNO=JNO+1
        IF(JNO.GT.NJ)RETURN
        GO TO 25
        END
        SUBROUTINE PRIMES(T,N,X,H)
        IMPLICIT REAL*B(A-H,O-Z)
        COMMON                    R(15),W(15),AKERN(15),XSUM,YSUM,
      1 E1SUM,E2SUM
      1,XPRI(15),YPRI(15),EPRI(15),AJPRI(76,15),UPRI(76)
        COMMON AUX(4),NAUX(11),NQUAD
        DIMENSION T(12610)
        REAL*8 JSUM
        DO 10 I=1,NQUAD
        IY=NQUAD+1
        XPRI(I)=T(IY)*XSUM
        IX=I
        YPRI(I)=-T(IY)/R(I)+T(IX)*YSUM
        IE=2*NQUAD
  10    EPRI(I)=-T(IE)/R(I)+T(IX)*(E1SUM+E2SUM)
        IF(N.EQ.(3*NQUAD))RETURN
        NJ=(N-3*NQUAD)/(NQUAD+1)
        JSUM=0.
        JNO=1
  15    DO 20 I=1,NQUAD
        IJ=(JNO-1)*(NQUAD+1)+3*NQUAD  +I
  20    JSUM=JSUM+T(IJ)*AKERN(I)
        DO 25 I=1,NQUAD
        IX=I
        IJ=(JNO-1)*(NQUAD+1)+3*NQUAD  +I
  25    AJPRI(JNO,I)=-T(IJ)/R(I)+T(IX)*JSUM
        UPRI(JNO)=JSUM*(E1SUM+E2SUM)
        JSUM=0.
        JNO=JNO+1
        IF(JNO.GT.NJ)RETURN
```

```
      GO TO 15
      END
      SUBROUTINE ESUM2(T,N,X,H)
      IMPLICIT REAL*8(A-H,O-Z)
      COMMON                R(15),W(15),AKERN(15),XSUM,YSUM,
     1 E1SUM,E2SUM
     1,XPRI(15),YPRI(15),EPRI(15),AJPRI(76,15),UPRI(76)
      COMMON AUX(4),NAUX(11),NQUAD
      DIMENSION T(12610)
      E2SUM=0.
      DO 10 I=1,NQUAD
      IEE=2*NQUAD  +I
   10 E2SUM=E2SUM+T(IEE)*AKERN(I)
      RETURN
      END
      SUBROUTINE XSUMS(T,N)
      IMPLICIT REAL*8(A-H,O-Z)
      COMMON                R(15),W(15),AKERN(15),XSUM,YSUM,
     1 E1SUM,E2SUM
     1,XPRI(15),YPRI(15),EP I(15),AJPRI(76,15),UPRI(76)
      COMMON AUX(4),NAUX(11),NQUAD
      DIMENSION T(12610)
      XSUM=0.
      DO 10 I=1,NQUAD
      II=I+NQUAD
   10 XSUM=XSUM+T(II)*AKERN(I)
      RETURN
      END
      SUBROUTINE YSUMS
      IMPLICIT REAL*8(A-H,O-Z)
      COMMON                R(15),W(15),AKERN(15),XSUM,YSUM,
     1 E1SUM,E2SUM
     1,XPRI(15),YPRI(15),EQRI(15),AJPRI(76,15),UPRI(76)
      COMMON AUX(4),NAUX(11),NQUAD
      YSUM=XSUM
      RETURN
      END
      SUBROUTINE KERNEL
      IMPLICIT REAL*8(A-H,O-Z)
      COMMON                R(15),W(15),AKERN(15),XSUM,YSUM,
     1 E1SUM,E2SUM
     1,XPRI(15),YPRI(15),EPRI(15),AJPRI(76,15),UPRI(76)
      COMMON AUX(4),NAUX(11),NQUAD
      DO 10 I=1,NQUAD
   10 AKERN(I)=R(I)*W(I)
      RETURN
      END
```

```
      SUBROUTINE ESUM1(X)
      IMPLICIT REAL*8(A-H,O-Z)
      COMMON                    R(15),W(15),AKERN(15),XSUM,YSUM,
     1 E1SUM,E2SUM
     1,XPRI(15),YPRI(15),EPRI(15),AJPRI(76,15),UPRI(76)
      COMMON AUX(4),NAUX(11),NQUAD
      E1SUM=DEXP(-(1.5-X))
      RETURN
      END
      SUBROUTINE RTSWTS
      IMPLICIT REAL*8(A-H,O-Z)
      COMMON                    R(15),W(15),AKERN(15),XSUM,YSUM,
     1 E1SUM,E2SUM
     1,XPRI(15),YPRI(15),EPRI(15),AJPRI(76,15),UPRI(76)
      COMMON AUX(4),NAUX(11),NQUAD
      DIMENSION RTS(13,15),WTS(13,15)
      IF(NQUAD.EQ.1)GO TO 20
      RTS(1,1)=.11270166
      RTS(1,2)=.5
      RTS(1,3)=.88729833
      WTS(1,1)=.27777777
      WTS(1,2)=.44444444
      WTS(1,3)=WTS(1,1)
      RTS(2,1)=.069431844
      RTS(2,2)=.33000948
      RTS(2,3)=.66999052
      RTS(2,4)=.93056815
      WTS(2,1)=.17392742
      WTS(2,2)=.32607258
      WTS(2,3)=WTS(2,2)
      WTS(2,4)-WTS(2,1)
      RTS(3,1)=4.6910077E-2
      RTS(3,2)=2.3076534E-1
      RTS(3,3)=.5
      RTS(3,4)=7.6923466E-1
      RTS(3,5)=9.5308992E-1
      WTS(3,1)=1.1846344E-1
      WTS(3,2)=2.3931433E-1
      WTS(3,3)=2.8444444E-1
      WTS(3,4)=WTS(3,2)
      WTS(3,5)=WTS(3,1)
      RTS(4,1)=3.3765242E-2
      RTS(4,2)=1.6939531E-1
      RTS(4,3)=3.8069041E-1
      RTS(4,4)=6.1930959E-1
      RTS(4,5)=8.3060469E-1
      RTS(4,6)=9.6623475E-1
```

```
WTS(4,1)=8.5662246E-2
WTS(4,2)=1.8038078E-1
WTS(4,3)=2.3395696E-1
WTS(4,4)=WTS(4,3)
WTS(4,5)=WTS(4,2)
WTS(4,6)=WTS(4,1)
RTS(5,1)=2.5446044E-2
RTS(5,2)=1.2923441E-1
RTS(5,3)=2.9707742E-1
RTS(5,4)=.5
RTS(5,5)=7.0292257E-1
RTS(5,6)=8.7076559E-1
RTS(5,7)=9.7455396E-1
WTS(5,1)=6.4742483E-2
WTS(5,2)=1.3985269E-1
WTS(5,3)=1.9091503E-1
WTS(5,4)=2.0897959E-1
WTS(5,5)=WTS(5,3)
WTS(5,6)=WTS(5,2)
WTS(5,7)=WTS(5,1)
RTS(6,1)=1.9855071E-2
RTS(6,2)=1.0166676E-1
RTS(6,3)=2.3723379E-1
RTS(6,4)=4.0828268E-1
RTS(6,5)=5.9171732E-1
RTS(6,6)=7.6276620E-1
RTS(6,7)=8.9833323E-1
RTS(6,8)=9.8014493E-1
WTS(6,1)=5.0614268E-1
WTS(6,2)=1.1119051E-1
WTS(6,3)=1.5685332E-1
WTS(6,4)=1.8134189E-1
WTS(6,5)=WTS(6,4)
WTS(6,6)=WTS(6,3)
WTS(6,7)=WTS(6,2)
WTS(6,8)=WTS(6,1)
RTS(7,1)=1.5919880E-2
RTS(7,2)=8.1984446E-2
RTS(7,3)=1.9331428E-1
RTS(7,4)=3.3787328E-1
RTS(7,5)=.5
RTS(7,6)=6.6212671E-1
RTS(7,7)=8.0668571E-1
RTS(7,8)=9.1801555E-1
RTS(7,9)=9.8408011E-1
WTS(7,1)=4.0637194E-2
WTS(7,2)=9.0324080E-2
```

```
WTS(7,3)=1.3030534E-1
WTS(7,4)=1.5617354E-1
WTS(7,5)=1.6511967E-1
WTS(7,6)=WTS(7,4)
WTS(7,7)=WTS(7,3)
WTS(7,8)=WTS(7,2)
WTS(7,9)=WTS(7,1)
RTS(8,1)=1.3046736E-2
RTS(8,2)=6.7468316E-2
RTS(8,3)=1.6029522E-1
RTS(8,4)=2.8330230E-1
RTS(8,5)=4.2556283E-1
RTS(8,6)=5.7443716E-1
RTS(8,7)=7.1669769E-1
RTS(8,8)=8.3970478E-1
RTS(8,9)=9.3253168E-1
RTS(8,10)=9.8695326E-1
WTS(8,1)=3.3335672E-2
WTS(8,2)=7.4725674E-2
WTS(8,3)=1.0954318E-1
WTS(8,4)=1.3463336E-1
WTS(8,5)=1.4776211E-1
WTS(8,6)=WTS(8,5)
WTS(8,7)=WTS(8,4)
WTS(8,8)=WTS(8,3)
WTS(8,9)=WTS(8,2)
WTS(8,10)=WTS(8,1)
RTS(9,1)=1.0885671E-2
RTS(9,2)=5.6468700E-2
RTS(9,3)=1.3492399E-1
RTS(9,4)=2.4045194E-1
RTS(9,5)=3.6522842E-1
RTS(9,6)=.5
RTS(9,7)=6.3477157E-1
RTS(9,8)=7.5954806E-1
RTS(9,9)=8.6507600E-1
RTS(9,10)=9.435312E-1
RTS(9,11)=9.891143E-1
WTS(9,1)=2.7834283E-2
WTS(9,2)=6.2790185E-2
WTS(9,3)=9.3145105E-2
WTS(9,4)=1.1659688E-1
WTS(9,5)=1.3140227E-1
WTS(9,6)=1.3 46254E-1
WTS(9,7)=WTS(9,5)
WTS(9,8)=WTS(9,4)
WTS(9,9)=WTS(9,3)
```

```
WTS(9,10)=WTS(9,2)
WTS(9,11)=WTS(9,1)
RTS(10,1)=9.2196828E-3
RTS(10,2)=4.7941371E-3
RTS(10,3)=1.1504866E-1
RTS(10,4)=2.0634102E-1
RTS(10,5)=3.1608425E-1
RTS(10,6)=4.3738329E-1
RTS(10,7)=5.6261670E-1
RTS(10,8)=6.8391574E-1
RTS(10,9)=7.9365897E-1
RTS(10,10)=8.849513E-1
RTS(10,11)=9.5205862E-1
RTS(10,12)=9.9078031E-1
WTS(10,1)=2.35876682E-2
WTS(10,2)=5.34696629E-2
WTS(10,3)=8.00391643E-2
WTS(10,4)=1.01583713E-1
WTS(10,5)=1.16746268E-1
WTS(10,6)=1.24573523E-1
WTS(10,7)=WTS(10,6)
WTS(10,8)=WTS(10,5)
WTS(10,9)=WTS(10,4)
WTS(10,10)=WTS(10,3)
WTS(10,11)=WTS(10,2)
WTS(10,12)=WTS(10,1)
RTS(11,1)=7.9084726E-3
RTS(11,2)=4.1200800E-2
RTS(11,3)=9.9210954E-2
RTS(11,4)=1.7882533E-1
RTS(11,5)=2.7575362E-1
RTS(11,6)=3.8477084E-1
RTS(11,7)=.5
RTS(11,8)=6.1522915E-1
RTS(11,9)=7.2424637E-1
RTS(11,10)=8.211746E-1
RTS(11,11)=9.007890E-1
RTS(11,12)=9.587992E-1
RTS(11,13)=9.920915E-1
WTS(11,1)=2.0242002E-2
WTS(11,2)=4.6060749E-2
WTS(11,3)=6.9436755E-2
WTS(11,4)=8.9072990E-2
WTS(11,5)=1.0390802E-1
WTS(11,6)=1.1314159E-1
WTS(11,7)=1.1627577E-1
WTS(11,8)=WTS(11,6)
```

```
     WTS(11,9)=WTS(11,5)
     WTS(11,10)=WTS(11,4)
     WTS(11,11)=WTS(11,3)
     WTS(11,12)=WTS(11,2)
     WTS(11,13)=WTS(11,1)
     RTS(12,1)=6.8580956E-3
     RTS(12,2)=3.5782558E-2
     RTS(12,3)=8.6399342E-2
     RTS(12,4)=1.5635355E-1
     RTS(12,5)=2.4237568E-1
     RTS(12,6)=3.4044382E-1
     RTS(12,7)=4.4597253E-1
     RTS(12,8)=5.5402747E-1
     RTS(12,9)=6.5955618E-1
     RTS(12,10)=7.576243E-1
     RTS(12,11)=8.436464E-1
     RTS(12,12)=9.136006E-1
     RTS(12,13)=9.6421744E-1
     RTS(12,14)=9.9314190E-1
     WTS(12,1)=1.7559730E-2
     WTS(12,2)=4.0079043E-2
     WTS(12,3)=6.0759285E-2
     WTS(12,4)=7.8601583E-2
     WTS(12,5)=9.2769199E-2
     WTS(12,6)=1.0259923E-1
     WTS(12,7)=1.0763192E-1
     WTS(12,8)=WTS(12,7)
     WTS(12,9)=WTS(12,6)
     WTS(12,10)=WTS(12,5)
     WTS(12,11)=WTS(12,4)
     WTS(12,12)=WTS(12,3)
     WTS(12,13)=WTS(12,2)
     WTS(12,14)=WTS(12,1)
     RTS(13,1)=6.0037409E-3
     RTS(13,2)=3.1363303E-2
     RTS(13,3)=7.5896708E-2
     RTS(13,4)=1.3779113E-1
     RTS(13,5)=2.1451391E-1
     RTS(13,6)=3.0292432E-1
     RTS(13,7)=3.9940295E-1
     RTS(13,8)=.5
     RTS(13,9)=6.0059704E-1
     RTS(13,10)=6.9707567E-1
     RTS(13,11)=7.8548608E-1
     RTS(13,12)=8.6220886E-1
     RTS(13,13)=9.2410329E-1
     RTS(13,14)=9.6863669E-1
```

```
      RTS(13,15)=9.9399625E-1
      RTS(13,1)=1.5376621E-2
      WTS(13,2)=3.5183023E-2
      WTS(13,3)=5.3579610E-2
      WTS(13,4)=6.9785338E-2
      WTS(13,5)=8.3134602E-2
      WTS(13,6)=9.3080500E-2
      WTS(13,7)=9.9215742E-2
      WTS(13,8)=1.0128912E-1
      WTS(13,9)=WTS(13,7)
      WTS(13,10)=WTS(13,6)
      WTS(13,11)=WTS(13,5)
      WTS(13,12)=WTS(13,4)
      WTS(13,13)=WTS(13,3)
      WTS(13,14)=WTS(13,2)
      WTS(13,15)=WTS(13,1)
      I=NQUAD-2
      DO 10 K=1,NQUAD
      R(K)=RTS(I,K)
   10 W(K)=WTS(I,K)
      RETURN
   20 R(1)=1.00
      W(1)=1.00
      RETURN
      END
      SUBROUTINE INT1(T,N,X,H)
      IMPLICIT REAL*8(A-H,O-Z)
      COMMON DUMMY(1310)
      COMMON           H2,H4,H24,RT
      COMMON N2,N3,N4,N5,N6,N7,N8,N9,NN,KFLAG,INDM,IDUM
      DIMENSION T(1)
C
C     CALC. CONSTANTS TO BE USED IN PROGRAM
      NN=N
      N2=N*2
      N3=N*3
      N4=N*4
      N5=N*5
      N6=N*6
      N7=N*7
      N8=N*8
      N9=N*9
      H2=H*0.5
      H4=H2*0.5
      H24=H/24.0
      RT=1./6.
C     CALC. Y PRIME FOR INITIAL CONDITIONS
```

```
        CALL DAUX(T,N,X,H)
        DO 1 I=1,N
        N91=I+N9
C     TEMPORARY STORAGE FOR Y
        N81=I+N8
        NNI=I+NN
        T(N91)=T(I)
C     STORE Y PRIME AT N-3 FOR USE IN A.M. INTEGRATION
      1 T(N8I)=T(NNI)
        KFLAG=0
        INDM=0
        RETURN
        END
        SUBROUTINE INT2(T,N,X,H)
        IMPLICIT REAL*8(A-H,O-Z)
        COMMON DUMMY(1310)
        COMMON        H2,H4,H24,RT
        COMMON N2,N3,N4,N5,N6,N7,N8,N9,NN,KFLAG,INDM,IDUM
        DIMENSION T(1)
C
C  INDM=FLAG FOR R.K. INTEGRATION WHEN EQUAL OR LESS THAN 3
C    KFLAG=FLAG FOR STORING PAST DERIVATIVES FOR USE IN A.M.
C      INTEGRATION
C
C                  RUNGE-KUTTA INTEGRATION (6 STEPS AT H/2)
C     NOTE- TWO STEPS OF R.K. ARE DONE WITH EACH CALL TO THIS
C           SUBR., SOTTHAT PRINTOUT POINTS WILL BE AT STEPS OF
C           H.
C
        INDM=INDM+1
C
C     STORE DERIVATIVES AS NEED FOR A.M.
        GO TO (10.11,9),KFLAG
C     STORE Y PRIME AT N-2
     10 DO 18 I=1,N
        N71=I+N7
        NNI=I+NN
     18 T(N7I)=T(NNI)
        GO TO 9
C     STORE Y PRIME AT N-1
     11 DO 19 I=1,N
        N61=I+N6
        NNI=I+NN
     19 T(N61)=T(NNI)
      9 CONTINUE
        GO TO 17
C           ADAMS-MOULTON INTEGRATION
```

```
C
C     STORE CURRENT Y AND Y PRIME IN TEMPORARY STORAGE
   12 DO 13 I=1,N
      N9I=I+N9
      N2I=I+N2
      NNI=I+NN
      T(N9I)=T(I)
   13 T(N2I)=T(NNI)
C     CALC. PREDICTED VALUE OF Y
      DO 14 I=1,N
      N2I=I+N2
      N6I=I+N6
      N7I=I+N7
      N8I=I+N8
      N9I=I+N9
      YP=T(N9I)+H24*(55.0*T(N2I)-59.0*T(N6I)+37.0*T(N7I)-
         9.0*T(N8I))
C     STORE AS PREDICTED FUNCTIONAL VALUE OF Y AT X=X+H
   14 T(I)=YP
C     STEP UP X
      X=X+H
C     CALC. Y PRIME USING PREDICTED Y
      CALL DAUX(T,N,X,H)
C     CALC. CORRECTED Y
      DO 15 I=1,N
      NNI=I+NN
      N2I=I+N2
      N6I=I+N6
      N7I=I+N7
      N9I=I+N9
      YC=T(N9I)+H24*(9.0*T(NNI)+19.0*T(N2I)-5.0*T(N6I)+T(N7I))
C     STORE AS NEW CURRENT VALUE OF Y
   15 T(I)=YC
C     CALC. Y PRIME TO BE USED IN NEW STEP
      CALL DAUX(T,N,X,H)
C     REARRANGE STORAGE OF PREVIOUS DERIVATIVES
      DO 16 I=1,N
      N8I=I+N8
      N7I=I+N7
      N6I=I+N6
      N2I=I+N2
C     Y PRIME (N-2) GOES TO (N-3)
      T(N8I)=T(N7I)
      KFLAG=KFLAG+1
      IF(INDM.GO.3)GO TO 12
      DO 9 K=1,2
C            CALC. K1
```

```
         DO 1 I=1,N
         N2I=I+N2
         NNI=I+NN
      1 T(N2I)=T(NNI)*H2
C          CALC. K2
C    STEP UP X
         X=X+H4
         DO 2 I=1,N
         N9I=I+N9
         N2I=I+N2
      2 T(I)=T(N9I)+0.5*T(N2I)
         CALL DAUX(T,N,X,H)
C    STORE K2
         DO 3 I=1,N
         N3I=I+N3
         NNI=I+NN
      3 T(N3I)=T(NNI)*H2
C          CALC. K3
         DO 4 I=1,N
         N9I=I+N9
         N3I=I+N3
      4 T(I)=T(N9I)+0.5*T(N3I)
         CALL DAUX(T,N,X,H)
C    STORE K3
         DO 5 I=1,N
         N4I=I+N4
         NNI=I+NN
      5 T(N4I)=T(NNI)*H2
C          CALC. K4
C    STEP UP X
         X=X+H4
         DO 6 I=1,N
         N9I=I+N9
         N4I=I+N4
      6 T(I)=T(N9I)+T(N4I)
         CALL DAUX(T,N,X,H)
C    STORE K4
         DO 7 I=1,N
         N5I=I+N5
         NNI=I+NN
      7 T(N5I)=T(NNI)*H2
C    CALC. PREDICTED VALUE OF Y
         DO 8 I=1,N
         N9I=I+N9
         N2I=I+N2
         N3I=I+N3
         N4I=I+N4
```

```
      N5I=I+N5
      Y=T(N9I)+RT*(T(N2I)+2.0*T(N3I)+2.0*T(N4I)+T(N5I))
C     STORE AS CURRENT VALUE OF Y
      T(I)=Y
C     STORE Y IN TEMPORARY STORAGE
    8 T(N9I)=T(I)
C     CALC. Y PRIME
      CALL DAUX(T,N,X,H)
C     Y PRIME (N-1) GOES TO (N-2)
      T(N7I)=T(N6I)
C     Y PRIME (N) GOES TO (N-1)
   16 T(N6I)=T(N2I)
   17 RETURN
      END
```

INDEX